The Usborne Introduction
to
Genes & DNA

Anna Claybourne

Designed and illustrated by
Stephen Moncrieff

Scientific consultant: Revd Professor Michael J. Reiss

Edited by Felicity Brooks
Additional designs by Laura Parker

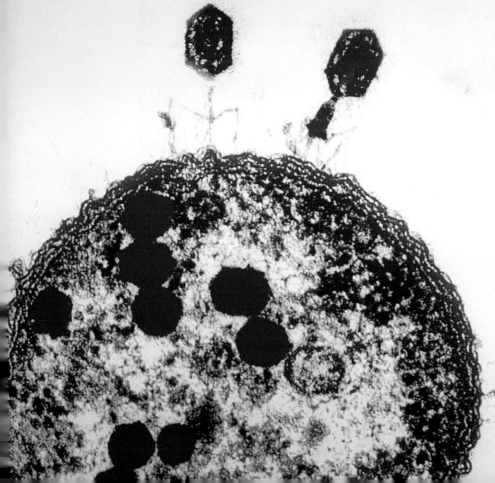

Viruses attack an
E. coli bacterium
by injecting it with
their DNA

Using internet links

This book contains descriptions of websites where you can find out more about genes and DNA. To visit the sites, go to the Usborne Quicklinks website at **www.usborne.com/quicklinks** and enter the keyword "genes".

Sometimes we add extra links too. Please note that the content of a website may change at any time and Usborne Publishing is not responsible for the content or availability of any website other than its own.

What you can do

Here are some of the things you can do at the websites we recommend:

• Zoom in on human DNA
• Experiment with cloning
• Find out about DNA testing in crime cases

Downloadable pictures

Pictures marked with a ★ in this book can be downloaded from the Usborne Quicklinks website and used in home or school projects. They must not be used for any commercial purpose.

Internet safety

When using the internet, children should follow our three basic rules:

• Always ask an adult's permission before using the internet.
• Never give out personal information, such as your name, address, telephone number or the name of your school.
• If a website asks you to type in your name or email address, check with an adult first.

For more information about using the internet, go to the Help and advice area at the Usborne Quicklinks website.

All living things, including humans, chimpanzees and plants, contain genes and DNA.

Site availability

The websites described in this book are regularly checked and reviewed by Usborne editors and the links at Usborne Quicklinks are updated. If a website closes down, we will replace it with a new link.

Contents

This science section explains what genes and DNA are and how they work. It can be difficult, but don't panic – even top scientists don't completely understand genes and DNA.

This section about passing on genes shows how genes and DNA run in families.

This is the history section, about how genes and DNA were discovered.

This section is about gene science today. It explores the gene stories that have hit the headlines – from cloned sheep to GM foods – and explains the facts behind them.

This section looks at "ethics" – the rights and wrongs of gene science. It explains why people are divided over whether cloning, GM foods and other inventions are good or bad ideas.

This is a useful reference section. It lists hundreds of important dates, names, numbers and gene science words.

The gene revolution

Genes and DNA are big news. Stories about cloning, GM foods, designer babies and DNA testing hit the headlines almost every day. But what exactly are genes and DNA? Where are they found? And why are they so important?

This sea lion, like all living things, is made of cells. Cells are controlled by the genes inside them. So genes are really instructions for controlling and running living things.

INTERNET LINKS

For links to websites where you can zoom into human DNA, go to **www.usborne.com/quicklinks**

Changing living things

In the last 50 years, gene scientists have gone beyond understanding genes and DNA, and learned to change or "genetically modify" them. This means they can change the way living things work and invent new varieties of animals and plants. Gene science is also behind many other new inventions and discoveries.

What are genes?

Genes are the instructions that make humans, animals, plants and other living things work. They are found inside the cells that make up all living things. Genes are made of a chemical called DNA. So "your genes" and "your DNA" often mean the same thing.

New discoveries

For a long time, scientists did not really know what made living things work. Genes and DNA have only really been understood in the last 100 years. Today, gene scientists know a lot about how genes work, and how they control cells and whole living things.

Gene worries

Some people are worried about the advances in gene science. They think it might be dangerous to alter genes and change living things. Campaigns are held to protest against some types of gene science.

A man dressed as a genetically modified animal protests against changing farm animals' genes.

Amazing inventions

These are some of the things gene scientists have done with their new knowledge of genes and DNA.

• **Mapping the genome** Scientists have made a map of the human genome. This is the complete set of genes needed to build and run a human being.

A tiny embryo (unborn baby) is held on the tip of a needle, ready to be tested for genetic diseases.

• **Cloning** Scientists have used gene science to make clones (exact copies) of many types of plants and animals.

• **Designer babies** Some diseases can be passed on from parents to children in genes. Doctors can help a couple to have a healthy child by checking cells for disease genes before the baby starts to grow.

A scientist takes samples from Ötzi, a mummified body found in the Alps. DNA testing can help archaeologists find out how old mummies are and what they looked like when they were alive.

• **Genetic Modification (GM)** This means changing the genes of a living thing to make it work differently. For example, scientists have created GM mice that glow in the dark.

• **Making medicines** Scientists have designed GM bacteria that can grow useful human body substances such as insulin, which is used to treat a disease called diabetes.

• **DNA fingerprinting** Because everyone has their own DNA, a "DNA fingerprinting" test can be used to track down a criminal, using a sample of skin or other body cells.

Understanding genes

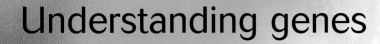

The next few pages explain the basics of what genes are, what they do, and how they work. Genes are complicated, and even gene scientists don't understand them completely. They have discovered a great deal, but there is still much more to find out.

A blue-ringed octopus looks and behaves the way it does because of its genes. They make its cells grow to form an octopus shape, and they make its body work so that it can live underwater.

How does life work?

For centuries, people have puzzled over some big questions about life:

• What gives a living thing its shape, size and colour?
• How are things like height and facial features passed on from parents to children?
• Why do all the members of a species look similar, but with slight differences?

Scientists now know that the answer to all these questions is "genes".

Genes and cells

Living things are made up of microscopic cells. A human, for example, has up to 100 trillion cells. A typical cell contains a nucleus, or control unit, with a set of genes inside. They control the cell by giving it instructions.

A typical human body cell

Nucleus containing genes that control the cell

Gene blueprints

Each species, or type, of living thing has its own special set of genes inside its cells. They make that species grow and work in its own special way. This is why humans, dogs and octopuses, for example, all look different — because they have different sets of genes.

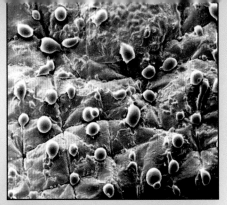

A photo of magnified human skin with drops of sweat. Skin and sweat are made of chemicals made by cells. Genes tell cells how to make them.

Recipe book

Cells don't use all their genes at once. Instead, the set of genes inside each cell is a bit like a recipe book. When a cell needs to do a job, it looks up the genes it needs and follows the instructions in them.

A chemical code

But how do genes store instructions? The answer is that they contain a code. Genes are made of DNA, and DNA stores instructions in the form of a pattern of four chemicals that act as code "letters". Cells follow a gene's instructions by reading this code.

The four colours in this diagram show the four chemicals DNA is made of. The way they are arranged acts as a code to store instructions.

★

This is a diagram of a piece of DNA, the substance genes are made of.

Differences

Although all humans have a set of human genes, we are not all the same as each other. For example, different people have different hair, eye colour and skin colour.

These differences are caused by slight differences in genes. Thanks to differences in the genes that make hair, it can be black, brown, fair, red, straight or curly.

These two children both have genes for making human hair and skin, but the recipes are slightly different.

Where are genes?

The pictures on these pages let you look inside a human cell, so you can see exactly where genes are and how they fit inside each cell. You can also see how genes are made of strands of DNA.

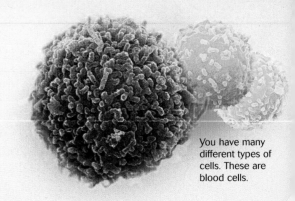

You have many different types of cells. These are blood cells.

Inside a cell

A typical human cell contains a nucleus and several smaller parts called organelles.

The nucleus is the cell's control unit.

Organelles do jobs for the cell, such as making and storing body chemicals.

Each cell is a self-contained unit, protected by a skin called the cell membrane. Inside, a cell has a control unit called the nucleus, and several other parts, called organelles.

Cells do all the jobs that keep the human body going, such as making body substances called proteins, and building new cells when old ones die. All the cells in your body work together to keep you alive and healthy.

The nucleus

This is a close-up of a cell nucleus.

Inside the nucleus are chromosomes, which contain DNA.

A cell's nucleus usually looks like a round ball near the middle of the cell. Genes are stored inside. Every cell nucleus in the body contains a copy of the same set of genes.

Genes are made of a chemical called DNA. DNA forms itself into long strands called chromosomes. Each nucleus contains 46 chromosomes, and the genes are arranged along the chromosomes.

This double spiral, linked by "rungs", is a strand of DNA.

Chromosomes

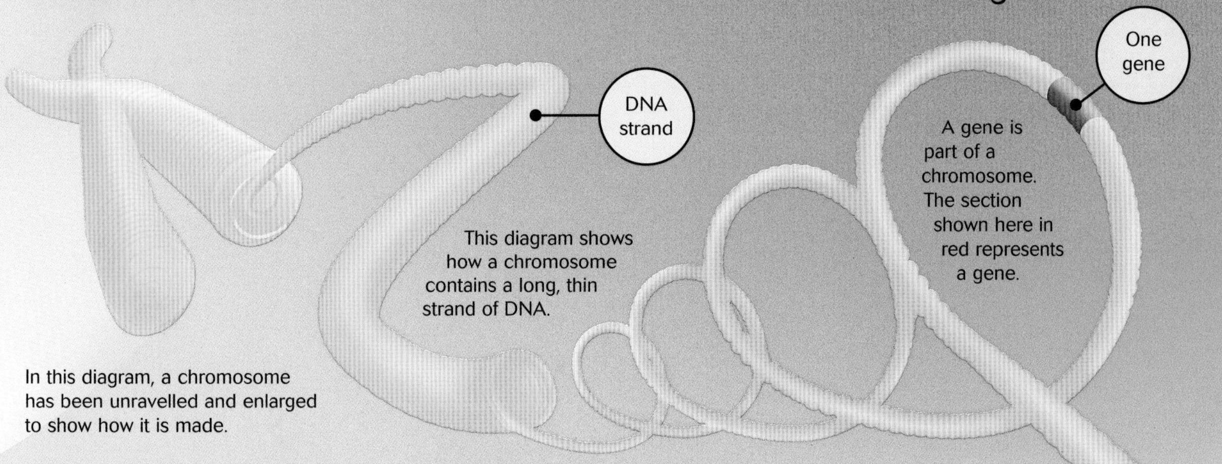

In this diagram, a chromosome has been unravelled and enlarged to show how it is made.

Chromosomes are long, thin strands of DNA, but they often coil themselves into short, thick shapes, like the ones in this diagram.

Together, the 46 chromosomes in each cell nucleus contain a full set of human genes. This means there's a complete set of about 20,000 human genes in almost every cell in your body. (A few unusual types of cells, such as red blood cells, do not have a nucleus and do not contain genes.)

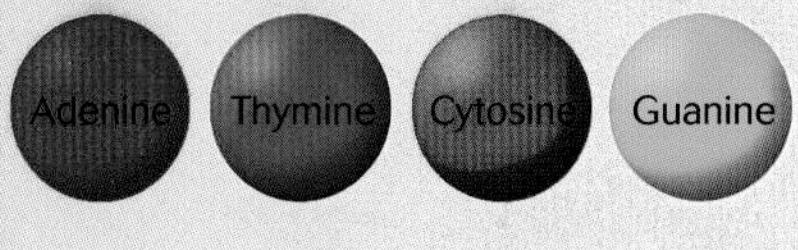

What is DNA?

DNA strand

This diagram shows how a chromosome contains a long, thin strand of DNA.

The letters DNA stand for deoxyribonucleic acid. This is a type of weak acid that naturally forms a long, twisted ladder shape called a double helix.

DNA contains four chemicals called bases. Their names are adenine, cytosine, guanine and thymine (or A, C, G, and T for short). They form pairs called base pairs. Each base pair makes one "rung" of the DNA ladder.

What is a gene?

One gene

A gene is part of a chromosome. The section shown here in red represents a gene.

A gene is a section of a chromosome, containing a special sequence of DNA. In a gene, the pattern of the four bases A, C, G and T acts as a code for making a particular body substance. There's more about how this works on pages 12-13.

The diagram below shows how the four bases are arranged in different patterns along the spiral DNA strand.

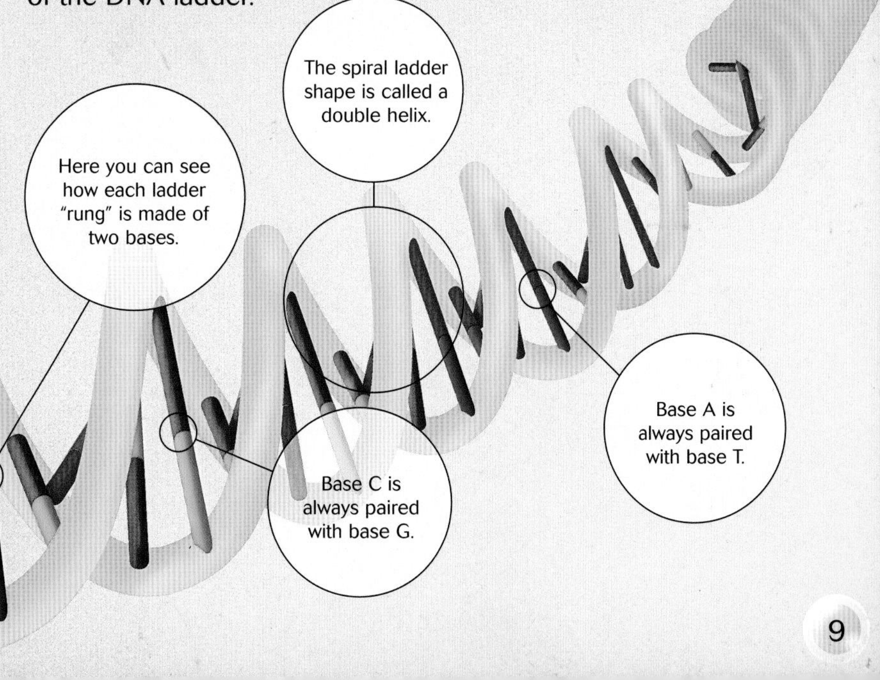

Base A   Base T   Base C   Base G

Adenine   Thymine   Cytosine   Guanine

Here you can see how each ladder "rung" is made of two bases.

The spiral ladder shape is called a double helix.

Base A is always paired with base T.

Base C is always paired with base G.

Chromosomes

Chromosomes are long strands of DNA. They are the storage units that hold genes. Having our DNA arranged into units like this makes it easier for us to pass genes on to the next generation. Almost every one of our cells contains 23 pairs of chromosomes, making 46 in each cell.

This picture, taken with a scanning electron microscope, shows some human chromosomes.

Chromosome pairs

You may have noticed that all the animals mentioned so far have even numbers of chromosomes. That's because chromosomes usually come in pairs. Each individual gets an equal set of chromosomes from each parent, making an even number in total.

Why 46?

No one knows why humans have 46 chromosomes. Different living things have different numbers, but it's not related to how big or complicated they are. For example, dogs have 78, tigers have 38, fruit flies have 8, and some ferns have 1,262.

Chromosomes aren't always the same length in different species. Salamanders have only 24 chromosomes, but these are so long that, altogether, a salamander has over ten times as much DNA as a human.

Large wild cats, such as the leopard above, and domestic cats, like the one below, are closely related. They both have 38 chromosomes in each cell.

Salamanders have 24 chromosomes.

12 from Mum

12 from Dad

Fruit flies have 8 chromosomes.

4 from Mum

4 from Dad

Are they X-shaped?

Chromosomes sometimes coil up into thick X shapes. You often see these in photos, because chromosomes are easiest to see when they are this shape. But a lot of the time, they are just long, thin strings of DNA floating in the cell nucleus. One scientist described them as being "like long strings of spaghetti in a fishbowl".

X and Y

Chromosomes are the same in men and women, except for one of the 23 pairs. One of the chromosomes in this last pair decides whether you will be male or female.

If you are female, both chromosomes in the pair are the same, and look like other chromosomes. But if you are male, one chromosome is shorter. It is called a Y chromosome.

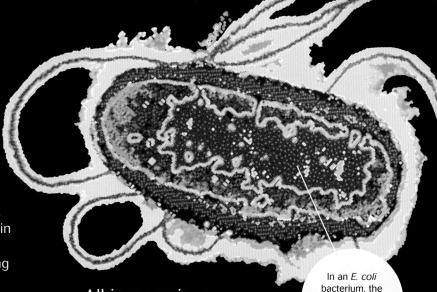

In an *E. coli* bacterium, the chromosome is a coiled-up ring of DNA (shown here in red).

All in one piece

Some simple living things, like bacteria, have their DNA all in one long string. An *E. coli* bacterium, for example, has all its DNA in a single, ring-shaped chromosome coiled up inside it.

Counting chromosomes

Scientists have named the human chromosomes according to their size. The biggest pair is called chromosome 1, the next biggest is called chromosome 2, and so on. The last pair is called XX or XY. This system helps scientists keep track of which genes are on which chromosomes.

This microscope photo shows a complete set of female human chromosomes. The colours have been added to the picture to make the chromosomes easier to see.

23rd pair of chromosomes ★

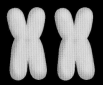

Women have two full-length chromosomes called X chromosomes.

Men have one full-length X chromosome, and one shorter one called a Y chromosome.

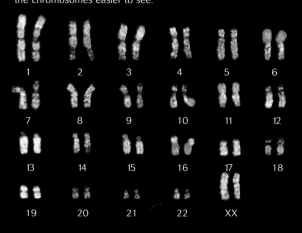

1	2	3	4	5	6
7	8	9	10	11	12
13	14	15	16	17	18
19	20	21	22	XX	

INTERNET LINKS

For links to websites where you can investigate chromosomes with online activities, go to **www.usborne.com/quicklinks**

The gene code

Genes are written in a code using the four bases, A, C, G and T. But how can a code of just four letters write a recipe for something as complicated as a human being?

Strings of bases

The diagram below shows part of a gene, made of a strand of DNA containing the four bases A, C, G and T. A cell follows the instructions in a gene by reading the pattern of bases along one side of the DNA strand.

Groups of three

The bases used to make a gene are arranged in groups of three. Each group of three bases works as a code. The four letters A, C, G and T can be arranged in 64 different groups of three, so there are 64 possibilities – for example ACG, ATT, TAC and GTA.

Below you can see how the bases that make up a gene can be divided into a series of three-letter codes.

An amino acid is a tiny molecule made up of atoms. This computer-generated image shows a molecule of the amino acid alanine.

This is an example of a three-letter code. It is made up of the sequence "TGA".

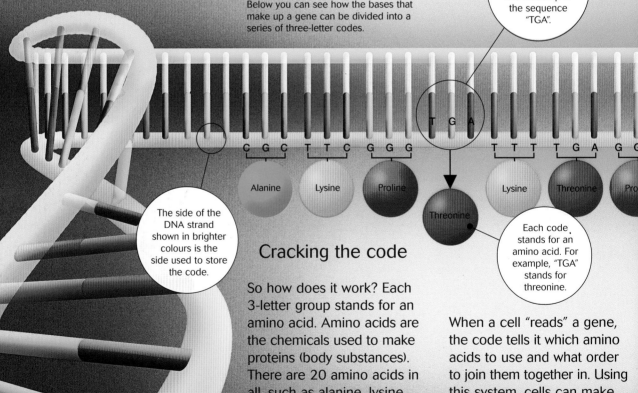

The side of the DNA strand shown in brighter colours is the side used to store the code.

Each code stands for an amino acid. For example, "TGA" stands for threonine.

T G A

C G C   T T C   G G G   T T T   T G A   G G

Alanine   Lysine   Proline   Lysine   Threonine   Pro

Threonine

Cracking the code

So how does it work? Each 3-letter group stands for an amino acid. Amino acids are the chemicals used to make proteins (body substances). There are 20 amino acids in all, such as alanine, lysine and proline. They come from food and are carried to your cells in your blood.

When a cell "reads" a gene, the code tells it which amino acids to use and what order to join them together in. Using this system, cells can make thousands of body proteins, each made up of a different arrangement of amino acids.

Doubling up

Since there are 64 possible 3-letter codes, but only 20 amino acids, some codes stand for the same thing. For example, as you can see in the diagram, the codes "TTT" and "TTC" both stand for the amino acid lysine.

Some of the codes have a different job. They act as start and stop signals, to mark where a gene begins and ends. This tells the cell where to start reading the code, and where to stop when the substance it is making is complete.

Junk DNA

Only about 1.5% of your DNA actually makes proteins. We still don't know what the rest of the DNA is for. It used to be called "junk DNA", and some scientists think that much of it serves no useful function but has simply accumulated over time.

This long sausage shape represents a chromosome containing several genes.

The red sections show which parts of the strand are genes.

The sections in between the genes may be junk DNA.

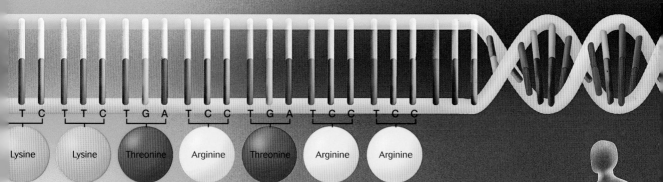

| T C | T T C | T G A | T C C | T G A | T C C | T C C |

| Lysine | Lysine | Threonine | Arginine | Threonine | Arginine | Arginine |

The same code

Almost all the world's living things use exactly the same gene code, made up of the same four letters. The same groups of three letters always stand for the same amino acids. This is why a gene can be taken out of one living thing and put into another, and it will still work.

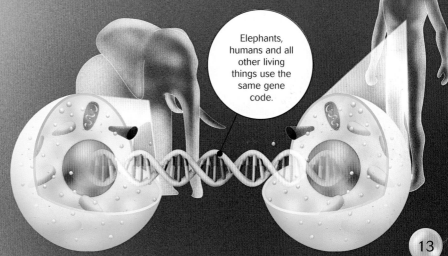

Elephants, humans and all other living things use the same gene code.

13

A fertilized human egg cell ready to grow into a baby

Building a baby

Everyone who has ever been born started life as a single cell. But how does a single cell know how to turn into a fully formed baby, with limbs, eyes, a brain, a heart, bones and skin, all made from different materials? The answer is in the genes the cell contains.

Blueprint for a baby

A baby normally begins when two cells, one from each parent, join together to make a new cell that can grow into a human. This is called a fertilized egg cell.

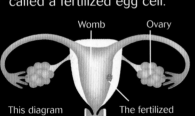

Womb   Ovary

This diagram shows the parts of a woman's body where a baby grows.

The fertilized egg cell attaches itself to the inside of the womb.

The fertilized egg cell contains a complete set of human genes. As the baby grows inside its mother's womb, the genes provide instructions for making all the body parts it needs. They decide its sex, eye and skin colour, and even the shape of its nose.

The embryo in this photo is about five weeks old. It already has a head, eyes, legs and arms, and the beginnings of fingers.

Making more cells

The cell starts dividing to make more cells just like it. These grow and divide too, until there is a cluster of several hundred cells called an embryo. Every time a new cell is made, it gets a complete copy of the first cell's set of genes.

Special cells

After a few days, some cells in the embryo start to follow special sets of gene instructions, and change into different kinds of cells. These "specialized" cells then form a range of different body parts.

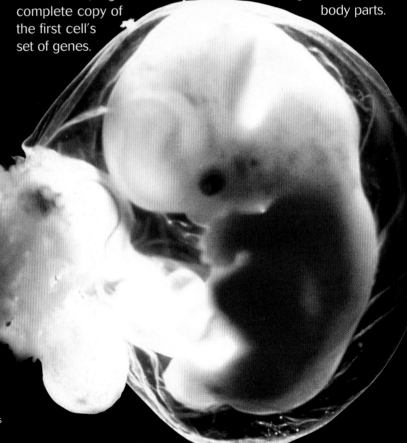

Bigger and bigger

Following the instructions in their genes, the cells keep growing and arranging themselves until, after about nine months, the baby is ready to be born. Instead of just one cell, it now has about two hundred million.

INTERNET LINKS

For links to websites where you can see embryos of living things and watch how they develop and grow, go to www.usborne.com/quicklinks

The finished baby

Finally, the baby is born; but the genes still have work to do. In the outside world, the baby keeps growing for another 20 years, still using instructions from its genes. Genes keep working throughout a person's lifetime, making new cells and all kinds of body substances.

Unusual shapes

Sometimes when a baby is growing in the womb, a mutation, or mistake, in its genes makes it grow in an abnormal way. It may grow extra limbs or toes, or unusually shaped body parts.

A gene mutation has given this frog a strange extra leg.

Different recipes

Other living things, from toads to tigers to trees, grow in the same way from a single cell. Each species has a different set of genes, which tell it to grow into its own particular shape.

As a chick develops, the genes inside its cells give them instructions for making claws, feathers, wings, a beak and all the other parts of a chick.

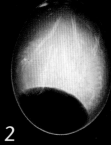

1

2

3

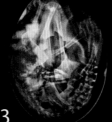

4

When it gets bigger and has a more human shape, the baby is known as a fetus. This fetus is about 5 months old.

Genes at work

Genes are an essential part of any living thing, and they get used every day. Cells can only make all the substances a body needs by following the instructions in their genes.

Making a protein

Whenever a living thing needs to make a body protein, it uses genes. For example, humans use a protein called insulin to digest sugar. When you need some insulin, your body sends a message to the cells in an organ called the pancreas, and they set to work. Follow these numbers to see what happens:

More proteins

Insulin is just one of thousands of proteins your cells can make. Here are some more:

Keratin is used to make hair, skin and nails.

Factor 8 is released to help your blood clot when you cut yourself.

Amylase helps you digest starchy foods such as potatoes.

Endorphins are made in your brain when you exercise. They can relieve pain or make you feel happy.

There are genes for every one of these substances. Whenever a cell needs to make one, it finds the right gene and follows the code.

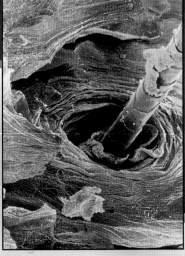

A microscope photo of a human hair. Skin and hair are made of the protein keratin.

Missing genes

Because cells rely on genes for instructions, it's vital that all your genes are in place and work properly. If a gene is missing or damaged, you might not be able to make a body substance you need.

For example, some people cannot make the blood-clotting protein factor 8, as their factor 8 genes are damaged. Without it, even small cuts do not heal. This disease is called haemophilia.

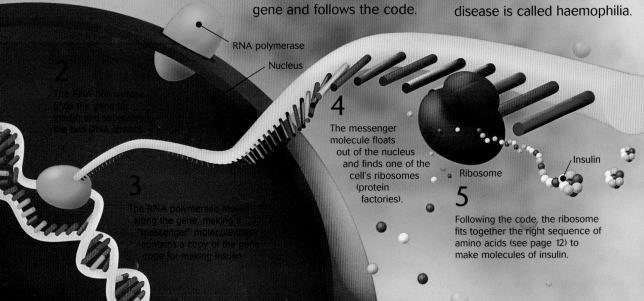

1 Inside a cell, a chemical called RNA polymerase goes into the cell nucleus.

RNA polymerase

Nucleus

2 The RNA polymerase finds the gene for insulin and separates the two DNA strands.

3 The RNA polymerase moves along the gene, making a "messenger" molecule that contains a copy of the gene code for making insulin.

4 The messenger molecule floats out of the nucleus and finds one of the cell's ribosomes (protein factories).

Ribosome

Insulin

5 Following the code, the ribosome fits together the right sequence of amino acids (see page 12) to make molecules of insulin.

DNA copies

Each new cell your body makes needs a copy of your chromosomes. The structure of DNA means chromosomes can copy themselves easily.

First, the DNA in a chromosome separates into its two strands, breaking apart the base pairs. Loose bases are always floating around inside cells. They join onto the strands, so that each half of the old chromosome builds itself up into a whole new chromosome.

A white blood cell dividing to make two new cells

Loose, unattached bases are attracted to the bases on the separated strands.

Gradually, as new bases are added, two copies of the original chromosome develop.

In the end, both the new pieces look just like the original piece of DNA.

Separated strands of original piece of DNA

Making new cells

Your body is constantly making new cells to help you grow, and to replace old, dead cells. Cells make new cells by splitting into two and making copies of their DNA. The way a cell grows, divides and separates into two is called mitosis. It works like this:

1
First, the whole cell grows bigger, making extra organelles and cytoplasm.

2
The cell's chromosomes copy themselves. The copies stay joined together in an X shape.

3
The X-shaped double chromosomes coil up and get shorter and thicker.

4
The X shapes separate into two sets of new chromosomes.

5
The nucleus divides into two, each with its own set of chromosomes.

6
The whole cell splits into two new cells, each with its own nucleus and DNA.

Passing on genes

All living things, including bacteria, plants, animals and people, can reproduce, or create new living things like themselves. When they reproduce, parents pass on their genes to their offspring. This is why babies – whether they're baby starfish, frogs or humans – grow up to look similar to their parents.

This arm was originally part of another starfish.

If a starfish loses an arm, the arm can sometimes grow itself a new body and become a new starfish.

Dividing

The simplest way to reproduce is to split into two. It's not only body cells that do this. Many single-celled creatures, such as bacteria, also reproduce by dividing themselves into two new "daughter" cells.

An *E. coli* bacterium grows until it is large enough to split in two.

DNA

The DNA in the bacterium copies itself to make two identical sets of genes, one for each new bacterium.

Budding

A few larger plants and animals can also reproduce alone, by "budding". They grow a smaller copy of themselves which eventually breaks free. Some worms, jellyfish and water animals called hydras (see page 40) can do this.

Babies made by budding have exactly the same set of genes as their parents. In other words, the baby is a "clone" – an exact copy – of the parent it came from.

It takes two

Most other animals, including humans, reproduce sexually. This means a cell from a male and a cell from a female join together to make a baby. The new baby is not an exact copy of one parent. Instead, it has half its mother's genes and half its father's genes.

These frogs are mating. Cells from both their bodies combine to make new cells which can grow into babies.

Reproductive cells

Cells that make new babies are called reproductive cells. In humans, male reproductive cells are called sperm, and female reproductive cells are called eggs. Unlike other cells, reproductive cells have only half a set of genes. Here you can see how they are made.

To make human reproductive cells, a cell divides twice. The new cells have only 23 chromosomes, not 46 as in a normal cell.

In this picture, hundreds of times bigger than life size, lots of sperm cells are surrounding an egg cell. One of them will join it and fertilize it.

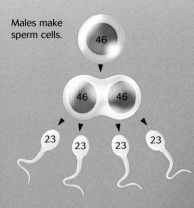

Females make egg cells.

Males make sperm cells.

When the cells join together, they make a fertilized egg cell. This has a full set of genes and can grow into a baby.

Sperm   Egg   Fertilized egg cell

Different every time

Reproductive cells are not all the same. As they are being made, some of the genes change places, so that each sperm or egg has its own special selection of half of a person's genes.

This means that if the same two parents have more than one child, the children will not have exactly the same genes as each other. Instead, each one will have a unique mixture of genes from their parents. This is why brothers and sisters can look different from each other.

Identical DNA

Identical twins are the only people who have the same genes as each other. They are made when the fertilized egg cell splits after the sperm and egg have joined. The separate parts grow into two babies with the exact same DNA.

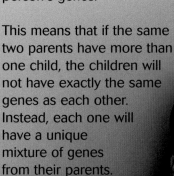

Most siblings (brothers and sisters) are not exactly alike. Each one has his or her own special set of genes.

Genetic traits

People often look like their parents, or even their grandparents. And as well as looks, you can inherit abilities, diseases and maybe even parts of your personality in your genes. Something that is inherited, or passed down from one generation to the next, in genes is called a genetic trait.

Genes and traits

Genetic traits, such as red hair and tallness, are caused by a particular gene or set of genes given to you by one or both of your parents. If you get a gene for a particular trait, it may or may not have an effect, depending on what other genes you have.

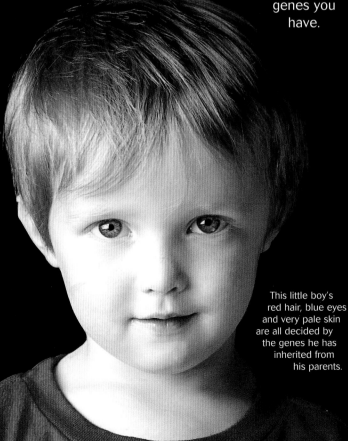

This little boy's red hair, blue eyes and very pale skin are all decided by the genes he has inherited from his parents.

Ear test

Do your earlobes dangle down, or are they attached directly to the sides of your head? This is an example of a genetic trait. Try looking at the earlobes of your family members to see which type they have.

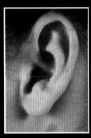 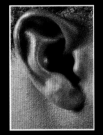

Dangling earlobe   Attached earlobe

How traits are passed on

Human cells contain 46 chromosomes. 23 of your chromosomes come from your mum, and the other 23 come from your dad. Each group of 23 contains a complete set of human genes. So everyone actually has two matching sets of chromosomes, and two copies of each gene.

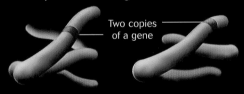

Two copies of a gene

Chromosomes from Mum   Chromosomes from Dad

The genes for earlobes exist in two different forms, or alleles. It's possible for one person to have both alleles. For example, you could have a dangling earlobe gene from your mum, and an attached earlobe gene from your dad.

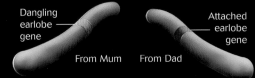

Dangling earlobe gene   Attached earlobe gene

From Mum   From Dad

Dominant genes

So which of the two genes decides what your earlobes look like? The answer is the dominant gene. With earlobes, the gene for dangling earlobes is dominant. This means it will always beat a gene for attached earlobes. As long as a person has one dangling gene, it will show up.

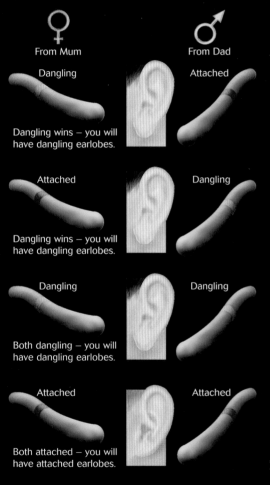

♀ From Mum   ♂ From Dad

Dangling — Attached

Dangling wins – you will have dangling earlobes.

Attached — Dangling

Dangling wins – you will have dangling earlobes.

Dangling — Dangling

Both dangling – you will have dangling earlobes.

Attached — Attached

Both attached – you will have attached earlobes.

This means a person could have attached earlobes, even if their parents don't. Both parents could have one dangling and one attached gene, giving them both dangling earlobes. But if they both pass on their attached genes to their child, he or she will have attached earlobes.

Recessive genes

The recessive gene is the name for the "weaker" gene which is beaten by a dominant gene. Recessive genes, such as the gene for attached earlobes, will only give a person their trait if they have two copies of the recessive gene.

Several diseases, such as cystic fibrosis and Tay-Sachs disease, are caused by recessive genes.

Nature and nurture

Genes are important, but they aren't the only things that make you who you are. People are also affected by their surroundings and the way they live their lives.

For example, genes can make someone tall, but a healthy diet makes people tall too. There are genes for being musical, but you still have to learn to play music. Many qualities like this are a mixture of "nature" (genes) and "nurture" (education and upbringing).

To become a good musician, you have to practise a lot.

Changing over time

Living things always pass on their genes to their offspring. That's why humans have baby humans, baby snakes hatch from snake eggs, and orange seeds become orange trees. But sometimes genes can change slightly as they are being copied.

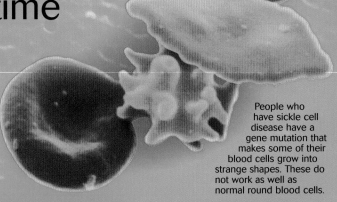

People who have sickle cell disease have a gene mutation that makes some of their blood cells grow into strange shapes. These do not work as well as normal round blood cells.

DNA mistakes

Every time a cell divides to make new cells, its DNA copies itself into the new cells. (You can see how this happens on page 17.)

This cell is dividing to make two new cells, each with a copy of the same DNA.

However, cells sometimes make mistakes as they copy the code – just like a typist making a mistake while copying text. These mistakes are called mutations.

Does it matter?

Often, mutations in DNA don't matter. They may affect only junk DNA, and not a gene. Many genes that do contain mistakes can still work normally. And sometimes, cells can repair mutations. But other times, a mutation can make a gene work differently or even make it stop working altogether.

Here is a diagram of a part of a gene. It is made up of three-letter groups. Each group stands for an amino acid that's needed to make a body protein.

Original three letters

Gene mutation

Below is the same sequence with a mistake in it. The mistake changes one of the three-letter groups into "ATC". Unfortunately, ATC is the code for "stop".

The gene will no longer work, because it will stop making the protein halfway through.

Passing it on

If a gene mutation happens when a cell copies itself inside your body, other cells will usually do the job of the damaged cell. But what if a mistake gets into an egg or sperm cell?

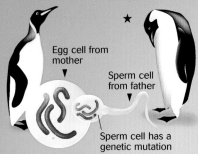

★

Egg cell from mother

Sperm cell from father

Sperm cell has a genetic mutation

Fertilized cell has genetic mutation.

Offspring has genetic mutation.

If the cell is used to make a baby, all that baby's cells will carry the same mistake. In this way, any plant or animal species can pass on gene mutations to its offspring.

Gene mutations have helped this species of plumed basilisk change over time to match its surroundings.

INTERNET LINKS

For links to websites where you can find out more about gene mutations and genes and evolution, go to **www.usborne.com/quicklinks**

Making changes

Gene mutations can happen at random, but some factors make them more likely.

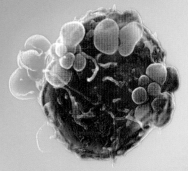

Gene mutations can cause cancer by making cells grow out of control. This picture shows a cancer cell.

Genes and evolution

Evolution is the name for the way species can change over time, thanks to useful gene mutations. Most scientists think all life evolved in this way, from simple, single-celled creatures to the many species we have today. This would explain why all living things have genes made up of the same basic DNA code.

Human gene sequences are very similar to those of other living things. About 98% of our DNA is the same as the DNA of chimpanzees.

Useful mistakes

Some gene mutations can be helpful. For example, imagine a species of yellow bugs living in a green forest. If one bug had a gene mutation that made it green instead of yellow, it would be harder for its enemies to see. It would probably survive longer than the others, have more babies, and pass on its "green" genes to them. They too would live longer and have more babies. Over time, the species would change from mostly yellow to mostly green.

For example, nuclear radiation causes extra gene mutations. This is why nuclear explosions can give people diseases. Sunlight can cause mutations in skin cells, which can sometimes lead to skin cancer.

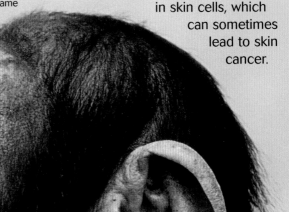

How genetics began

The study of genes and DNA is called genetics. Gene scientists, called geneticists, have only recently begun to understand exactly how genes work. But even before we understood them, genes played a very important part in human culture, and people have always had theories about them.

A statue of Tutankhamun, a king of ancient Egypt. DNA testing has been used to find out which other royals he was related to.

On the farm

Farmers have been carrying out a kind of genetics known as selective breeding for thousands of years. By selecting only the biggest or best animals and plants for breeding and sowing new crops, they allowed only the most useful genes to be passed on.

A 12,000-year-old cave painting of a cow-like animal. Cows were among the first creatures to be bred by farmers.

Birth and breeding

Many ancient peoples saw that children resembled their parents. Up to 2,000 years ago, early Hindus realized that diseases could run in families. Most early societies also had kings and queens who passed on their power to their children. This showed that people believed qualities such as royalty or "noble blood" were passed on from parents to children.

Greek genetics

The ancient Greek scientist Aristotle thought babies got all their qualities from their father. But Hippocrates, an early doctor, was closer to the truth. He said men's and women's bodies contained liquids that mixed to make a baby. The liquids fought each other to decide which parent's features got passed on.

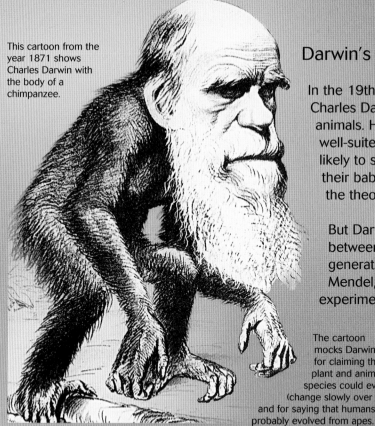

This cartoon from the year 1871 shows Charles Darwin with the body of a chimpanzee.

The cartoon mocks Darwin for claiming that plant and animal species could evolve (change slowly over time) and for saying that humans probably evolved from apes.

Darwin's discoveries

In the 19th century, the English naturalist Charles Darwin studied differences between animals. He saw that animals that were well-suited to their surroundings were more likely to survive and pass their qualities to their babies. This was the starting point for the theory of evolution (see page 23).

But Darwin didn't know *how* differences between animals were passed from one generation to the next. However, Gregor Mendel, an Austrian monk who had been experimenting on pea plants, had the answer.

INTERNET LINKS

For links to websites where you can watch animations about Mendel's experiments and find out how they worked, go to www.usborne.com/quicklinks

Mendel's peas

Mendel bred pairs of pea plants with different qualities, such as different heights or different-shaped peas. Instead of being a blend of both parents, the baby plants either did or did not inherit each quality. From this, Mendel saw that there must be "factors" or "atoms" of inheritance – now called genes. Plants had the qualities they had because of the genes they inherited.

Sadly, no one was interested in Mendel's results and Darwin did not understand their meaning. They weren't recognized as important until long after Mendel's death.

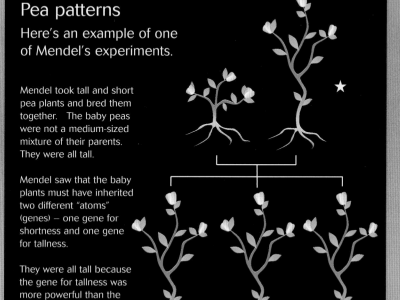

Pea patterns
Here's an example of one of Mendel's experiments.

Mendel took tall and short pea plants and bred them together. The baby peas were not a medium-sized mixture of their parents. They were all tall.

Mendel saw that the baby plants must have inherited two different "atoms" (genes) – one gene for shortness and one gene for tallness.

They were all tall because the gene for tallness was more powerful than the shortness gene. In modern terms, it was a dominant gene.

Discovering DNA

When Gregor Mendel discovered genes, he knew something must be passing on genetic traits. But he couldn't actually see genes, because they were too small. In the 20th century, new technology finally allowed scientists to work out where genes and DNA were and what they looked like.

INTERNET LINKS

For links to websites about early microscopes, the achievements of famous gene scientists and the discovery of DNA, go to **www.usborne.com/quicklinks**

Smaller and smaller

Over the past 400 years, we have been able to look at smaller and smaller objects, as more powerful microscopes have been developed.

In the 1800s, Joseph Lister made new lenses which could magnify up to 1,200 times. This 1828 cartoon shows a lady looking in horror at the germs (shown as tiny monsters) that could now be seen in drinking water.

Today's powerful microscopes can magnify up to a million times and show their results on a computer screen. This picture, made using a scanning electron microscope, shows human chromosomes.

Left: In the 1660s, Robert Hooke worked with early microscopes that allowed him to see fleas at about 200 times life size. This is one of his flea drawings.

Micro-discoveries

Microscopes were invented in the 17th century, but the early ones were not very powerful. It took many years to develop better microscope design and stronger lenses. By the late 1800s, microscopes were so good that scientists could look inside cells.

They saw chromosomes inside the nucleus doubling and splitting when new cells were made. They realized that qualities could be passed on from parents to children in chromosomes.

Mendel was right!

In the years leading up to 1900, three scientists named de Vries, Correns and von Tschermak came up with the same results as Mendel had 40 years earlier. They also saw that "atoms of inheritance" – or genes – must be in chromosomes, and that they came in pairs.

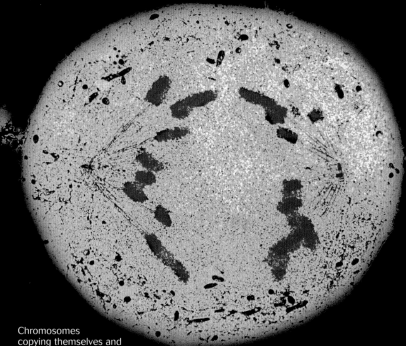

Chromosomes copying themselves and separating before a cell divides. When scientists saw this, they realized that chromosomes might contain genes.

Genetics takes off

Now, genetics was on the right track, and new discoveries were made quickly. Scientists worked out that each chromosome was a string of many genes. They also realized that out of the 23 pairs of chromosomes in a human, one pair decided whether a baby would be a boy or a girl (see page 11).

In the 1940s, two biologists, George Beadle and Edward Tatum, discovered that different genes coded for different body proteins. And another biologist, Oswald Avery, found that chromosomes and genes were made of a substance named deoxyribonucleic acid – also known as DNA.

Franklin's pictures

But what did DNA look like? In 1950, a chemist named Rosalind Franklin developed a way of using X-rays to photograph tiny objects.

The patterns in Franklin's photos showed scientists that DNA had a spiral shape.

Franklin took pictures of the DNA molecule which showed that it was shaped like a spiral, or helix.

A Nobel Prize

One of Franklin's colleagues, Maurice Wilkins, gave a copy of her results to two other scientists, James Watson and Francis Crick. Using models of the four chemical bases that make up DNA, they were then able to work out that the spiral was a double helix. They also discovered how the DNA code worked, and how DNA could copy itself (see page 17).

In 1962, Watson, Crick and Wilkins were awarded a Nobel Prize for their work on DNA. Franklin might have shared the prize too, but she had died of cancer in 1958.

This modern computer-generated model shows the shape of a molecule of DNA.

Gene science today

In the 21st century, genetics has become one of the most important of all sciences. Geneticists are finding the genes that contribute to diseases, making new medicines and redesigning living things. The different branches of gene science are changing all our lives.

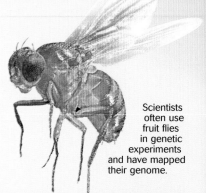

Scientists often use fruit flies in genetic experiments and have mapped their genome.

DNA testing

It's now possible to read the patterns in a person's DNA. DNA testing has many uses:
• to see if you have genes that carry genetic diseases.
• to see whether two people belong to the same family.
• to trace a criminal using a sample of hair or skin from a crime scene.
• to find out more about mummies and preserved prehistoric animal and plant life.

Genome mapping

Genome mapping means working out the genome (complete DNA sequence) of a living thing. Humans, *E. coli* bacteria, fruit flies and many other species have now had their genomes mapped.

Genome mapping doesn't actually show how genes work. It just provides a long sequence of DNA code for the species being mapped. This raw data can then be used to find out what particular genes do and how they work. For example, scientists study the human genome to find genes that cause diseases.

To map a genome, cells are taken from a living thing such as a worm or fruit fly.

The chromosomes containing the DNA are extracted from the cells.

A C T A G G C   A C T A A G C

The DNA is analyzed and powerful computers work out its sequence.

Genetic engineering

Genetic engineering means making changes to DNA in order to change the way living things work. It is used to create new types of crops and farm animals, and to make bacteria that can make medicines. In theory, humans can be genetically engineered too.

These chickens have been genetically engineered to have no feathers.

The Ancient of Days, painted by William Blake in 1794, shows a god-like creator similar to the Christian God.

Ethical debates

Many people feel very strongly about gene science. While some believe it should be used to develop useful inventions, others are opposed to interfering with the way living things work. Scientists have been accused of "playing God" by creating new life forms. Arguments about the ethics (rights and wrongs) of genetics are a big part of the world of modern science.

INTERNET LINKS

For links to websites where you can find out about gene science today, go to www.usborne.com/quicklinks

Big business

Gene science is not just important scientifically – it's important for business too. Many genetic developments, such as bug-resistant crops and gene therapy medicines, can make a lot of money. Powerful biotechnology companies employ their own geneticists, or pay university scientists, to work on gene research that could lead to money-making inventions.

Scientists have created genetically engineered salmon that grow much faster than normal salmon. They could make big profits for fish farmers.

Changing ourselves

Soon, we may begin to alter our own genes. This could have all kinds of medical uses. For example, it could help us to eliminate genetic diseases, or live for much longer than we can now.

We might eventually be able to do other things too, such as changing our appearance or improving our memory, intelligence and strength.

The human genome

In February 2001, scientists announced that they had finished a first map of the human genome, the complete sequence of human DNA. This knowledge will help us find out which genes do which jobs – information which is vital to medicine.

INTERNET LINKS

For links to websites where you can discover more about the mapping of the human genome, go to
www.usborne.com/quicklinks

The genome race

In 1990, an international group of scientists launched the Human Genome Project – a plan to make a map of a complete set of human genes by the year 2010. The results would be available for anyone to use. But in 1998, an American gene scientist named Craig Venter started a private company called Celera Genomics. Using a new computer system to read DNA sequences, he planned to map the genome himself and patent the results so that people would have to pay to use them. The two organizations began a race to finish the genome map. In the end, it was a tie. The two sides came to an agreement in the year 2000, and finally made a joint announcement that the map was complete.

Left: the American genome scientist and businessman Craig Venter. Behind him is a computer display of groups, or "colonies", of genes waiting to be copied and processed as part of the complex genome mapping process.

A computer display showing sequences of bases in human DNA. The four colours stand for the four bases in the gene code.

How was it done?

The map was made using cells taken from a group of volunteers and from Venter himself. To read the DNA code, scientists used a method called gel electrophoresis, in which DNA is separated into pieces in gel-filled glass tubes. Powerful computers analyzed the results to give the sequence of bases in each gene.

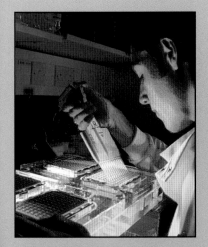

A Human Genome Project scientist preparing human DNA for mapping at the Sanger Centre in Cambridge, UK

Gaps in the map

Although the announcement was made in 2001, at the time the map of the human genome was not quite complete. Many areas of DNA are made up of short, repeated sequences which were very hard for sequencing machines to read. Analysis and interpretation of all the data collected continues to this day.

Maps and patents

Besides the human genome, scientists continue to map the genomes of many other species, such as worms, fruit flies, dogs, thalecress (a kind of weed) and rice. As well as being useful in itself, mapping other genomes helps us to understand more about the human genome, because all living things share some of the same genes and DNA patterns.

Some companies, including Celera, have tried to patent gene sequences. This means they would have a private licence and other scientists would have to pay to use them.

This is a close-up microscope photo of a fruit fly. Fruit flies are very important in genetics, because they breed fast and scientists can observe how their genome changes over time.

Facts and figures

• The human genome has a sequence of 3.2 billion base pairs. This would fill 200 telephone directories. If you read it out loud, it would take you more than 50 years.

• There are about 20,000 different genes in the human genome.

• If you took all the DNA from a single human cell, joined it together and stretched it out, it would be about 2m (6ft) long. All the DNA in your body would reach to the Sun and back 600 times.

Genetic engineering

Genetic engineering means changing the genes of living things. It has thousands of possible uses, from GM foods to amazing new materials, new medicines and even completely new species. However, it is still not fully understood, and no one knows whether it is completely safe.

These E. coli bacteria have been genetically engineered so that they produce human insulin.

Why do it?

Living things grow the way they do because of instructions in their genes. So, for example, an E. coli bacterium follows instructions in its genes to grow into a sausage shape and make the proteins it needs to stay alive.

By changing those genes, scientists can make E. coli behave differently. For example, if they insert a gene for human insulin into an E. coli bacterium, the E. coli will make lots of insulin, which they can collect and use. Many other living things can be genetically altered too.

A genetically engineered glowing mouse

How it works

Genetic engineering is usually done by taking a living cell from one species and adding a gene from another species. These pictures show how scientists made a genetically engineered glowing mouse.

— Jellyfish cell

— Jellyfish DNA

— Gene for glowing protein

1 This jellyfish has a gene that makes a glowing protein. This makes the jellyfish glow in some types of light.

★

Virus DNA —

Virus —

— Inserted jellyfish gene

2 The glowing gene is taken from a jellyfish cell and spliced (inserted) into a virus. (The virus has been modified so it cannot carry diseases.)

— Mouse cell

— Mouse DNA

Viruses work by inserting their DNA into a cell.

3 The genetically engineered virus attaches itself to the fertilized mouse egg cell.

4 The virus delivers the glowing gene into the egg cell nucleus, where it joins the mouse DNA.

5 The genetically engineered mouse egg grows into an adult mouse which will make the glowing protein. The glow is too faint to see under normal lights, but it can be detected using a special camera.

What could it do?

Glowing mice have only been developed as an experiment, but there are millions of more practical uses for genetic engineering.

However, there are problems. Many people think changing living things in this way is morally wrong. And genetic engineering could have unexpected effects, or could be used in a harmful way. For example, someone might create genetically engineered, super-deadly bacteria to use as a weapon.

This is anthrax, a germ that has been used as a weapon. Genetic engineering could make it even more dangerous.

More uses

Here are some genetic engineering inventions. Some of these are in use already, and others are being developed.

• **Killer moth** To reduce the number of crop-eating caterpillars, scientists have released GM moths designed to pass on deadly disease genes to their relatives.

• **Organ donor** Scientists may be able to engineer pigs to grow human organs for use in transplants.

• **GM sheep** Sheep can be made to produce human proteins in their milk to help treat diseases such as cystic fibrosis and emphysema.

• **Killer cotton** A gene from insect-killing bacteria has been added to cotton plants, to make them poisonous to the insect pests that feed on them.

Genetic engineering will let scientists make large amounts of strong, flexible spider silk.

• **Spider rope** A gene from a spider has been inserted into some goats. Their milk now contains tiny strands of spider silk which can be made into a strong and stretchy rope.

• **Supermaize** By changing gene sequences in maize, scientists have made a new type of maize containing extra nutrients.

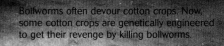

Bollworms often devour cotton crops. Now, some cotton crops are genetically engineered to get their revenge by killing bollworms.

GM foods

Of all the things genetic engineering can do, GM (genetically modified) foods are probably the most talked about. GM foods could help to end hunger around the world, but many people worry about how safe they are.

People often imagine GM foods as weird combinations of different species. In fact, they usually look exactly like normal food.

"Frankenfoods"

GM foods have been called "Frankenfoods" after the scientist in Mary Shelley's novel *Frankenstein*. In the book, Frankenstein creates a monster that turns against him and attacks his family.

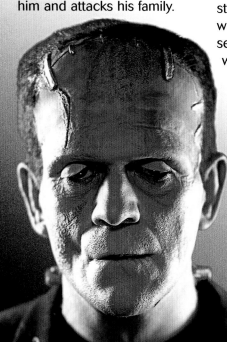

Some protesters see GM foods as man-made monstrosities, like Frankenstein's monster. This photo shows Boris Karloff as Frankenstein's monster in a 1935 film.

On your table

Millions of us have already eaten GM foods. Some GM tomatoes, for example, have had their genes altered to stop them from going soft while they are growing. For several years they were widely sold in tomato paste.

The GM foods we eat have all been tested for safety. But many people are still afraid that they could be bad for us in ways we do not yet understand. Because of protests against GM foods, many shops have stopped selling them.

Fishy strawberries

These diagrams show how one type of GM food, a strawberry that resists frost damage, is made.

The flounder is a fish that lives in icy seas. It has a gene that stops it from freezing to death. Strawberries are soft fruits that can easily be damaged by frost.

1 The flounder's antifreeze gene is copied and inserted into a small ring of DNA taken from a bacterium.

Bacterium's DNA

Antifreeze gene

2 The DNA ring containing the flounder gene is put into a second bacterium.

Second bacterium

3 This second bacterium is used to infect the strawberry cell. The flounder's antifreeze gene enters the strawberry's DNA.

Strawberry cell

Antifreeze gene

4 The new GM strawberry cell is grown into a GM strawberry plant, which can be bred many times.

Thanks to their new gene, the GM strawberries make a protein which helps them resist frost. They don't contain any other fish genes, and do not taste or smell of fish.

More fears

There are other reasons why some people worry about GM foods. One is that new combinations of genes might lead to unexpected changes. A carrot modified to grow bigger, for example, might also be poisonous. Growing GM crops could harm the environment too. For example, crops modified to be poisonous to pests could make some insect species extinct.

The two tomatoes on the left have been genetically altered to resist mould.

Blowing in the wind

When crops are growing, their pollen can blow away and land in other fields. In this way, pollen from GM crops, containing modified genes, can combine with non-GM crops and alter them too. Growers try to keep GM crops separate from non-GM crops, but it doesn't always work. If GM foods do turn out to be harmful, this will be a big problem.

This anti-GM foods protester is pulling up GM crops growing in a field in England, and sealing them inside a plastic bag.

One study suggests that some GM maize crops can kill monarch butterflies.

Fighting famine

There are millions of people in the world who don't have enough to eat. This is often because their crops are eaten by pests, or fail due to poor soil or lack of rain. GM crops designed to resist pests or survive in thin, dry soil might help to solve these problems and prevent famines.

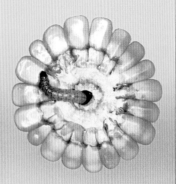

GM maize can be designed to resist pests such as this corn borer.

Genetic medicine

Gene science is about to revolutionize the way we treat and prevent diseases. Scientists can now find genes that cause illnesses, and are learning how to repair or replace them. They can use genetically modified bacteria to make medicines more cheaply and safely, and soon they may even be able to make whole human organs.

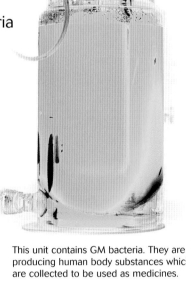

This unit contains GM bacteria. They are producing human body substances which are collected to be used as medicines.

Making medicines

Until recently, medicines had to be collected from plants or animals, or made using chemicals. Now, scientists are making body substances, such as human blood-clotting factor and insulin, by using genetically engineered bacteria to grow them.

This tray holds samples of white blood cells taken from a sick patient. They are being genetically altered to make them better at fighting disease.

Making body parts

Organ transplants can go wrong if the body rejects the new organ. Gene scientists may be able to solve this problem by growing new cells, tissues, and maybe whole new organs, from cells cloned from the patient. They would have the same DNA as the patient, so they would not be rejected.

It may also be possible to add human genes to an animal such as a pig, so that it grows organs perfect for transplanting into a human patient.

Screen for a gene

With DNA testing, doctors can now find out how likely you are to get diseases such as cancer or Huntington's disease. As therapies improve, some genetic diseases could be cured before they even begin.

These two chromosomes belong to a patient who only has one copy of a gene (shown in red) that protects against cancer, instead of two. This shows she has an increased risk of getting cancer.

A cure for cancer?

Cancer happens when body cells grow out of control. Scientists have found a gene called p-53 which normally keeps cells under control. They think that in some people with cancer, the disease begins because the p-53 gene doesn't work properly – perhaps because of a mistake in the gene code.

Traditional cancer treatments, such as chemotherapy, often have bad side-effects. So experts are now looking for ways to cure cancer by modifying faulty DNA to make the p-53 gene work.

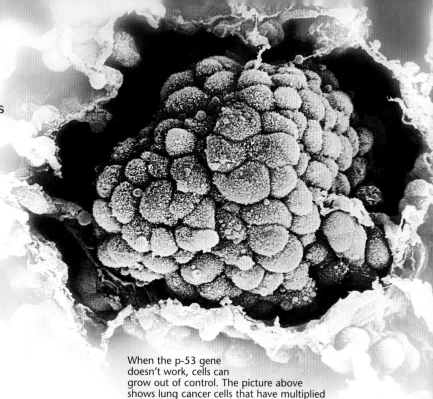

When the p-53 gene doesn't work, cells can grow out of control. The picture above shows lung cancer cells that have multiplied and formed a lump, known as a tumour.

Gene therapy

Gene therapy means repairing or replacing genes that cause diseases. This technique is quite new, but it is beginning to work. The pictures here show how gene therapy may one day be used to treat cystic fibrosis.

Cystic fibrosis sufferers have a gene mutation which means their cells cannot make a protein that the lungs need. Their lungs fill with sticky mucus, making it hard to breathe.

In this magnified picture, you can see a lump of mucus (the yellow part) inside the lung of a cystic fibrosis patient.

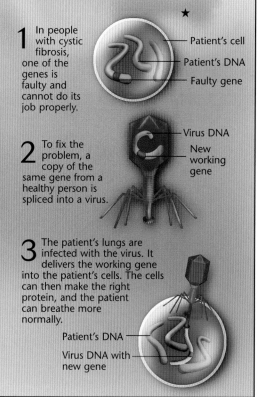

1 In people with cystic fibrosis, one of the genes is faulty and cannot do its job properly.

★
Patient's cell
Patient's DNA
Faulty gene

2 To fix the problem, a copy of the same gene from a healthy person is spliced into a virus.

Virus DNA
New working gene

3 The patient's lungs are infected with the virus. It delivers the working gene into the patient's cells. The cells can then make the right protein, and the patient can breathe more normally.

Patient's DNA
Virus DNA with new gene

Designer babies

Instead of using gene science to cure diseases, why not just pick babies who won't get those diseases at all? This is already happening, and the result is a "designer" baby – a baby selected before birth for its healthy genes.

Genetic diseases

To get a genetic disease such as cystic fibrosis (CF), a baby has to inherit two copies of the disease gene from its parents. If both parents carry one disease gene and one healthy gene, there's a one-in-four chance the baby will get two disease genes and end up with the disease.

However, if parents know their children will be at risk, they can use a method called Pre-implantation Genetic Diagnosis (PGD) to ensure they have a healthy baby.

A healthy fetus (unborn baby) growing in its mother's womb

A human egg cell, shown about 600 times bigger than life size, being held on the end of a pipette ready to be injected with a sperm cell.

How it works

This is how PGD is used to help a couple with CF genes to have a healthy child.

First, the mother takes a drug to make her body produce extra egg cells. The egg cells are collected in a glass dish and fertilized with sperm cells from the father.

Each embryo can develop into a baby, but about one in four of them will have CF. The embryos are tested for CF using DNA testing methods (see page 44).

When they are ready to be tested, the embryos have only eight cells each.

★

Sperm

Egg

When a sperm fertilizes an egg cell, it becomes an embryo which can grow into a baby.

One cell is removed to be used in the test. This doesn't harm the embryo.

Two or three healthy embryos are selected and implanted in the mother's womb. One or more of them may grow into a baby.

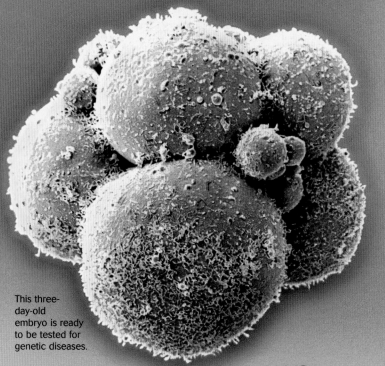

This three-day-old embryo is ready to be tested for genetic diseases.

Is it right?

There has been some debate about PGD, because the unwanted embryos are thrown away. Some people say that an embryo is a sacred human life and should not be discarded for having a disease.

Boy or girl?

As well as testing for diseases, PGD can show whether an embryo is male or female. This means it's possible to choose the sex of a baby. There can be health reasons for choosing a boy or a girl. For example, the genetic blood disease haemophilia only affects boys. Parents can avoid passing it on by only having girls.

However, many families prefer boys for cultural reasons. If they could all choose, there would be a world-wide girl shortage.

Families using PGD often have twins or triplets, because more than one embryo is implanted. This scan shows a pair of twins in the womb.

Baby to the rescue

In the year 2000, a couple in the USA used designer baby technology in a new way. Their daughter Molly had a genetic disease. She could be cured by a cell transplant from a healthy baby whose genes closely matched hers. So the couple used PGD to have a healthy boy, Adam. Cells from his umbilical cord were used to treat Molly.

Molly Nash and her doctor with baby Adam, who was "designed" to help Molly survive

What next?

New knowledge about the human genome (see page 30) might allow us to identify genes for all kinds of qualities, such as looks, height and intelligence. Will people start choosing – or even engineering – embryos for these qualities?

Many people are worried that designing babies in this way is wrong, or could cause problems. There's more about this on pages 48-49.

Cloning

Cloning means making copies of living things by copying their DNA. For a long time, it was associated with sci-fi stories, but today, cloning is a scientific reality.

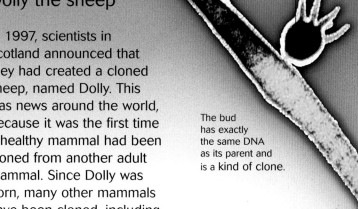

This hydra reproduces by growing a "bud" (a small version of itself) from its side. The bud grows into a new hydra and eventually breaks free.

Bud

The bud has exactly the same DNA as its parent and is a kind of clone.

What is a clone?

A clone is a living thing that is an exact genetic copy of another living thing. Some creatures, like the hydra, clone themselves naturally. You can clone a plant by taking a cutting from it. And there have always been humans with the exact same DNA – identical twins. But deliberate scientific cloning only began recently.

Cloned humans would look alike because they would have exactly the same DNA.

Dolly the sheep

In 1997, scientists in Scotland announced that they had created a cloned sheep, named Dolly. This was news around the world, because it was the first time a healthy mammal had been cloned from another adult mammal. Since Dolly was born, many other mammals have been cloned, including cats, deer, dogs, horses, mice and monkeys.

Human cloning

Some scientists think that the method used to clone Dolly could work on humans too. Some people want to clone humans as a way of having babies, or even as a way to "replace" a child who has died. Some want to try it just to see if it will work.

However, most countries have banned human "reproductive" cloning – using cloning to make human babies. Many people think it's wrong to create new life artificially, or to make a child who has no choice about being a clone.

How cloning works

These pictures show how a human could be cloned. Instead of combining DNA from two parents (see page 19), cloning involves copying the DNA of just one parent.

1 Scientists take an egg cell from a woman and remove its DNA.

[Egg] cell with DNA removed

DNA

2 Then they take a complete cell, such as a skin cell, from another person's body.

Skin cell with complete set of DNA

3 The empty egg cell and the complete skin cell are placed close together and given an electric shock which makes them fuse together.

4 The new cell, containing DNA from just one person, is implanted into a woman's womb, where it begins to grow.

5 Nine months later, the cloned baby would be born.

INTERNET LINKS

For links to websites where you can experiment with cloning and find out more about Dolly the sheep, go to **www.usborne.com/quicklinks**

Identical twins are natural clones. A cloned baby would be like an identical twin of the person it was cloned from, but younger.

Stem cell cloning

Stem cells are cells found in embryos that can turn into any kind of body cell. They can be used to repair damage to body organs. It's now possible to make stem cells that match a patient's DNA by making a cloned embryo from that patient. Debates are raging about whether doing this is ethical.

The future of cloning

In the future, cloning could become a convenient way to breed farm animals. It could also help scientists to breed identical animals for research. And although it is illegal in most countries, some scientists are already working on human cloning. One day, some parents might have their children as a result of cloning.

Dolly, the world-famous cloned sheep, was born in 1996 and introduced to the world in 1997.

Living for ever

The oldest human beings on record have lived to be about 120. Most people die before reaching 100. But what makes us grow old and die? The answer is, at least partly, in our DNA. People are now beginning to hope that by making changes to our DNA, we could live for much longer – or even for ever.

This bone scan shows the leg bone of an elderly woman with osteoporosis. This disease happens when cells stop making some types of hormones as people get older, making bones weak and brittle. The weak areas are shown in pink and green.

DNA time-bomb

As we live our lives, our cells keep dividing to replace old, dead cells. But cells can only divide so many times. As they slow down and stop, our bodies grow less able to repair damage and more at risk of disease. This is how we get old.

Ageing is partly caused by telomeres – repeated sections of DNA on the ends of chromosomes. Each time a cell divides, its telomeres wear down. When they get too short, a cell cannot divide any more.

In a telomere (the end section of a chromosome), the same sequence of six bases is repeated thousands of times.

Repair worker

Some cells in the body, such as sperm and egg cells (see page 19) and stem cells (see page 41), contain a protein called telomerase. It repairs telomeres so that they don't get any shorter.

With genetic engineering, it may be possible to alter our cells so that they all make telomerase. In theory, this might stop our cells from wearing out, and slow down the ageing of our organs.

INTERNET LINKS

For links to websites where you can find out more about life expectancy and how we grow old, go to
www.usborne.com/quicklinks

Is living for ever a good idea?

Throughout history, people have dreamed of living for ever and told each other legends about defeating death. But would everlasting life really be a good thing?

Why live for ever?
• Some people want to live for ever simply because they are afraid of death.

• If we could keep brilliant scientists and inventors alive for ever, they could keep contributing to progress.

• If we lived for ever – or for much longer than we do now – we might look after our environment better, because we would live long enough to see the results of our actions.

Why not?
• Many countries already suffer from overcrowding and food shortages. If people didn't die, these problems would get worse and worse.

• If you were immortal, you could never retire, because your pension would run out. You'd have to work for ever.

• Many religious people believe they will become immortal in another world after they die. Because of this, they have no need or desire to stay on Earth for ever.

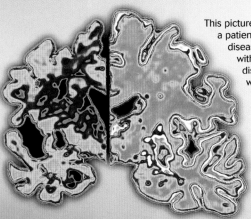

This picture shows a scan of the brain of a patient suffering from Alzheimer's disease (on the left) compared with a healthy brain. Alzheimer's disease makes brain cells weaken and die in old age.

A scientist freezes cells taken from a baby's umbilical cord, to be used in research into the uses of stem cells.

Replacement parts

Although research into telomeres is at an early stage, a few experts have suggested that gene science could eventually help us to live much longer lives. One possibility would be to have a sample of your cells collected at a very young age. They would be copied and used to make cloned stem cells containing your DNA. These could be made into body tissues to make new organs for you when you are older.

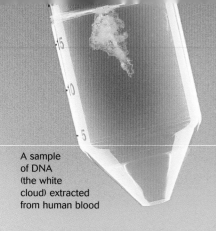

A sample of DNA (the white cloud) extracted from human blood

DNA testing

Everyone, apart from identical twins, has a unique pattern of DNA. So DNA testing is a very accurate way of identifying people. Genetic tests can also find out if people are related, show whether people will get certain diseases, and reveal the age of ancient bodies

DNA fingerprinting

Testing someone's DNA to discover their identity is called DNA fingerprinting. It is used to match suspects to evidence in crime cases, and can also prove people's innocence. In the USA, at least 18 convicts on death row have been released after DNA tests proved they were not guilty after all.

See for yourself

The results on the right are real DNA test results from a real-life crime scene. You can see for yourself that the attacker's DNA only matches the DNA of suspect 1. The pictures below show how DNA is tested.

1 ★

Detectives take samples of skin, blood or other body cells from the crime scene, and from the suspects and the victim.

2

Long strands of DNA are extracted from the cells and cut up into smaller pieces.

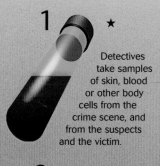

3

The DNA pieces are processed by putting them in a special gel. Different people's DNA forms different patterns in the gel. The patterns are then transferred to a piece of film to be analyzed.

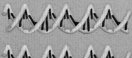

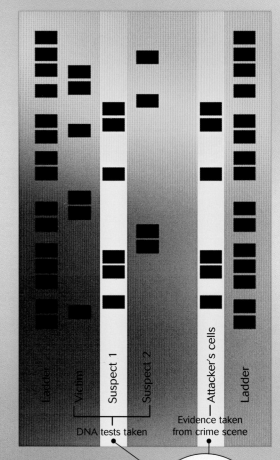

Ladder

Victim

Suspect 1

Suspect 2

Attacker's cells

Ladder

DNA tests taken

Evidence taken from crime scene

4

In this case, the final results show DNA patterns of the victim, two suspects, and some skin cells left by the attacker. The "ladder" lines are control lines for measuring the other marks against.

These two traces match, showing that suspect 1 and the attacker who left the cells behind are likely to be the same person.

Testing for diseases

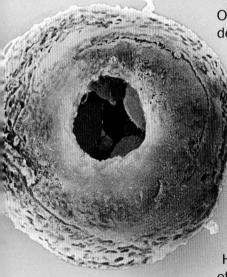

This is a very small embryo, or unborn baby, with just a few cells. One cell has been removed for testing to find out if the embryo has disease genes.

Testing for disease genes is similar to DNA fingerprinting. It can find out if an unborn baby is likely to get certain illnesses. Adults can also be tested to see if they are likely to have diseases such as heart disease. But gene tests could be used against you too. Employers or life insurance companies might turn you down if they knew you would become ill. So some countries now have laws saying people can keep their genetic details private.

INTERNET LINKS

For links to websites where you can investigate DNA testing and its uses in crime cases, identifying diseases, and archaeology, go to **www.usborne. com/quicklinks**

All on file?

One day, all our genetic details could be stored in enormous databases. Computers could match DNA from crime scenes to offenders in minutes. Medical researchers could study a gene database along with medical information to find out which genes and which diseases go together.

However, because of the risk of the information falling into the wrong hands or being misused, some people are opposed to DNA databases.

DNA of the dead

Archaeologists are now using DNA testing too. By comparing DNA from a mummified body with the DNA from another mummified body, they can work out if the two individuals were related.

This 2,000-year-old woman was found in a peat bog in Denmark. Peat preserves body tissues, making it possible to take DNA samples from some bog mummies.

Amber (fossilized tree sap) sometimes contains small trapped animals, such as this spider. By testing DNA from them, scientists can find out more about prehistoric creatures.

Right or wrong?

Genetic science has already led to many useful inventions. However, some people think that some of the things gene scientists can do, and some of the methods they use, are ethically wrong.

This hairless rat has been bred especially for use in laboratory tests. It would never have had a life at all otherwise. Is this right or wrong?

What are ethics?

Ethics are moral guidelines that help us decide what is right and wrong. They apply to lots of things, not just genetics. For example, is it right to help a terminally ill person to die if they want to? Is it right to test medicines on animals? Is it right to put criminals to death? Ethical debates like these are going on all the time.

These protesters are invading a field of GM crops to show that they think growing GM crops is wrong.

Ethics and genetics

There is a huge amount of ethical debate about genetic science. This is largely because it deals with living things. Many people feel that creating new life is a matter for religion, not science. Some worry that animals, and maybe humans, will suffer if they are involved in genetic experiments. Others think it may be unethical not to use genetic science if it can improve people's health, living standards and quality of life.

INTERNET LINKS

For links to websites where you can explore the ethics of genetic science and some of the arguments for and against its uses, go to www.usborne.com/quicklinks

Big questions

Here are some of the main questions in genetic ethics:

• **Playing God** Is it right to make changes to living things and create new types of creatures?

• **Evil uses** Should genetic engineering be banned in case it ends up being used in an evil or dangerous way?

• **Cloned babies** Should human cloning be allowed so that childless couples can have babies? Is it ethically right to clone animals?

• **Design for life** Should parents be allowed to choose the sex of their child? What about other qualities, such as height, eye colour and intelligence?

• **Access to DNA** Who does a human DNA sequence belong to – the scientist who discovered it, or the person it came from? Or does it belong to us all?

• **Stopping suffering** Despite the potential problems, can it be wrong to use gene science, if it might cure illnesses, prevent disability or end famines?

• **DNA testing** Should people be tested in advance to see if they will get a genetic disease? Should insurance companies and employers be allowed to see the results?

Cloning technology could be used to grow organs for transplants.

All parents want their babies to be as healthy and happy as this one. But some people argue that using gene science to prevent diseases and disabilities could rob us of genetic variety and undermine the right to be different.

On the Web

Many people who feel strongly about gene science use sites on the World Wide Web to express their views. Remember that whenever you find a website about genes and DNA, you could be reading just one side of a complex argument.

Perfect people

Today, designer babies are only selected to avoid genetic diseases. But one day, parents might choose to give their babies genes for looks and intelligence too. In theory, this could create a beautiful, clever human race.

INTERNET LINKS

For links to websites where you can find out about eugenics and "designer babies", go to **www. usborne.com/quicklinks**

As scientists learn more about genes, people may be able to select qualities such as eye colour, height and intelligence for their babies.

Changing ourselves

There are plenty of ways of trying to improve ourselves. Some people have cosmetic surgery to alter their faces and bodies, and most of us can learn new skills. But altering our genes is more serious, especially as any changes could be passed on to our children.

Who decides?

"Good" and "bad" qualities are a matter of opinion. For example, you might think it's good to be clever, but a dictator might prefer you to be easy to control. There's a risk that whoever has the most money and power could impose their idea of perfection on others.

Eugenics

Trying to "improve" the genes of a human population is called eugenics, and it is not new. In the 1930s, the German Nazi government tried to wipe out what it saw as "bad" qualities — including being Jewish, disabled or a Roma gypsy. First, the Nazis tried to stop these groups from having children. Later, they simply killed them, in what became known as the Holocaust.

The Nazis killed their victims in special camps. These men were rescued from one in 1945.

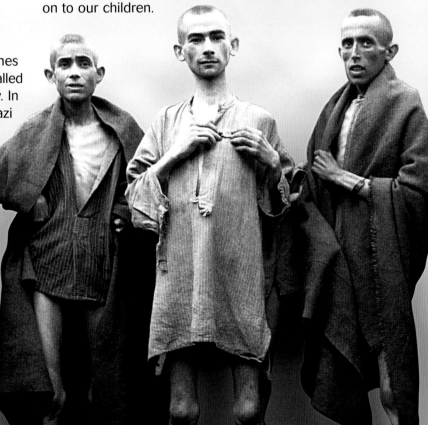

A genetic elite

If the technology for designing "perfect" people does become available, it will probably be expensive. Would that mean that only people with plenty of money could "design" themselves to be cleverer and more attractive?

This picture is taken from the 1997 film GATTACA, which explores a future world where only genetically perfect people, known as "valids", are allowed to have the best jobs. The hero, Vincent, is not a valid but pretends to be one so he can get a job as an astronaut.

The great scientist Einstein (right) had dyslexia, which makes reading and writing difficult. If his parents had been able to avoid having a baby with dyslexia by using designer baby technology, he might never have been born.

Imperfect heroes

Most people agree that it's a good idea to save children from genetic conditions such as cystic fibrosis. On the other hand, selecting embryos to avoid "bad" genes could rob us of other, more useful qualities too.

Many great achievers have had genetic conditions. If we had only picked "perfect" embryos in the past, many people who have contributed a lot to society would never have been born.

Don't forget nurture!

Whatever genes you have, nurture – that is, lifestyle and the way you are brought up – also has a huge effect on who you are and what you can do. Diet, exercise and education are at least as important as your genes in achieving success and happiness.

Making money

Gene science is not just about finding out facts. Our new knowledge leads to new inventions, which can be sold to make a profit. Big businesses put vast amounts of money into gene science research, so they can make even more money out of the inventions and discoveries that follow.

Who and how?

There are various ways to profit from gene science. Companies such as Celera Genomics decode plant and animal genomes and sell the results. Companies such as Monsanto and Nexia have developed GM plants, or new substances such as BioSteel (made with milk from genetically modified goats). Drug companies design gene therapy methods and genetic tests, which they sell to hospitals.

Business boost

Business has helped gene science to move forward at an amazingly fast rate. Many genetics companies employ top scientists to work on their projects. Some also team up with genetics departments in universities. They provide the university scientists with plenty of money, in order to support research into their shared areas of interest.

Cows are being genetically engineered to make useful medicines in their milk. This GM bull is being raised to breed a herd of genetically engineered cows.

Good or bad?

Some scientists are happy to accept money from business. Universities are often short of cash, and a big business grant can be vital to keep a department running – especially since genetic science requires expensive precision equipment.

A scientist's gloved hand holds a tiny genetically engineered plant seedling during experiments to develop new GM crops. Crops that can resist cold and disease, or produce extra-large yields, can make big profits for their developers.

Other scientists say that accepting money means they are not free to do their own research. Also, some people worry that scientists working for powerful genetics companies might cut corners and compromise on safety in order to make more money.

Patenting genes

A patent is a licence for an invention. It lets you make money from your idea and stops other people from copying it. Some companies have now started to patent gene sequences they have discovered or designed.

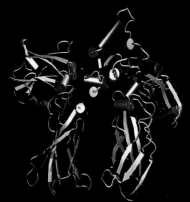

This is a computer model of a protein called erythropoietin. The gene that makes it has been patented.

This has caused an angry debate. Opponents of gene patenting argue that genes aren't inventions, but information that we all share. Others say genes belong to whoever they came from.

Geneticists are keen to study the genes of Native Americans, such as these Navajo girls, to find out why some Native Americans are resistant to certain diseases.

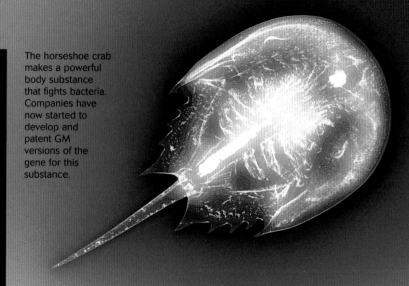

The horseshoe crab makes a powerful body substance that fights bacteria. Companies have now started to develop and patent GM versions of the gene for this substance.

Gene gold-rush

Now that genes can be patented, many companies are "prospecting" for genes that could be useful in medicine or industry. They are sampling thousands of plants and animals in order to find and patent as many genes as possible. They also sample genes from people. The government of Iceland decided to collect genetic data from all its citizens and store them in a database which can be sold to genetics companies.

Making monsters

Genetic engineering already allows us to create new life forms, and the technology is constantly improving. Is it possible that a dangerous new species, such as a killer bacterium, might be made accidentally? Or could genetic engineering be deliberately used to create deadly new weapons?

The legendary chimera had a lion's head, a serpent for a tail, and a pair of wings on its back.

Accidental killer

Scientists have already made a killer virus by accident. Luckily for humans, it only kills mice. In 2001, Australian researchers engineered a virus as part of a vaccine to stop mice from having babies. To their surprise, the GM virus killed all the mice.

New GM viruses and bacteria are being developed all the time, so an accident affecting humans could happen. If a deadly GM germ escaped from the lab, it might be almost impossible to get back.

Viruses work by breaking into cells and injecting their genes into them. The virus below, called a bacteriophage, is attacking a bacterial cell.

Fear of the chimera

The chimera is a monster from Greek mythology – part serpent, part lion and part goat. Protesters have used the chimera as a symbol of the dangers of genetic engineering, claiming that irresponsible scientists could make monstrous new creatures. These animals could be dangerous, or they might suffer as a result of being created.

Today, gene scientists usually take an existing species and add a single gene from another. But as science progresses, chimera-like species could be made. Human and animal DNA could even be combined to make a "half-human" species.

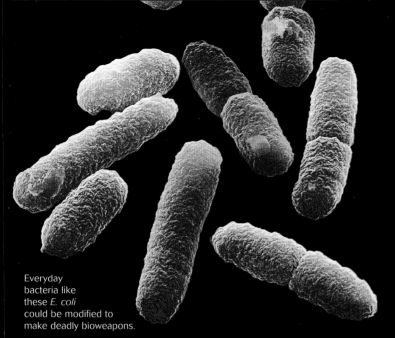

Everyday
bacteria like
these *E. coli*
could be modified to
make deadly bioweapons.

GM bioweapons

Bioweapons (short for biological weapons) are designed to harm their victims by poisoning them or making them sick. Weapons such as poisonous chemical gases have been used for years. But new gene technology could make bioweapons more deadly than ever.

Many countries are researching genetically engineered weapons. Some experts think GM germs could have already been used as weapons, for example in the Gulf War of 1991.

Suits like this are designed to protect people from bioweapons.

INTERNET LINKS

For links to websites where you can find out about potential dangers of genetic engineering, go to **www. usborne.com/quicklinks**

Dangerous science

It's obvious that genetic science could lead to some nasty accidents, and that in the wrong hands it could also be used destructively. However, this is also true of other sciences. Almost everything humans have ever invented – from the wheel to the Web – can be used in both good and bad ways.

Nuclear fission, developed in the 20th century, is a useful energy source. But it also has destructive power in the form of bombs.

Since so much gene technology is already available, most experts believe that genetic research should not be banned. Instead, they say governments should pass laws regulating genetic engineering, with safety guidelines for scientists to follow.

Into the future

Where will genetics lead? No one knows for certain. Most people agree that the advance of genetics is a huge scientific revolution which will change our lives for ever – but they can't agree on exactly how. Here are just a few of the possibilities that have been suggested.

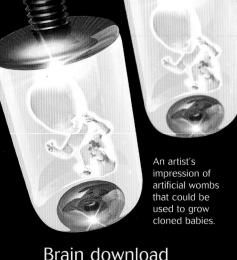

An artist's impression of artificial wombs that could be used to grow cloned babies.

INTERNET LINKS

For links to websites where you can explore possible advances in the future of genetic science, go to
www.usborne.com/quicklinks

New species

As genetic engineering progresses, scientists could create new species, instead of just changing existing ones. They might be able to write a whole new genome on a computer, piece together the DNA sequence and grow the new species as a clone. For example, they could make a new bacteria able to clean up oil spills.

Clone culture

One day, cloning might be a normal way to have a baby. Some people even predict that to avoid the discomfort of pregnancy and childbirth, cloned human babies will be grown inside other animals, such as cows, or in computer-operated artificial wombs. Many clones could be made from the same DNA, so some people might have hundreds of "twins".

Animal cloning might be a way to save endangered species, such as the white rhino.

Brain download

Some people hope that cloning will help them live for ever. When they get old, they plan to clone themselves and "download" their personality into their new body. There is currently no way to do this, but there are people alive today who believe it will be invented in time for them.

Disaster ahead?

Some people believe that improved medicines, better crops and new inventions that result from genetics could make life a lot happier for billions of people around the world.

Others disagree. They predict that genetic science could wipe out life on Earth. This could be caused by a killer GM virus or bacterium. Or a biological imbalance could be created by modifying plant and animal DNA.

Information about DNA has been sent into space as a message to any other intelligent life that may exist.

DNA profiling

As DNA testing improves, DNA profiling will become common. Even when there is no suspect, police may be able to analyze DNA taken from a crime scene, and use the genes to find out if the criminal is male or female, and what he or she looks like. Here are some of the features they may be able to discover:

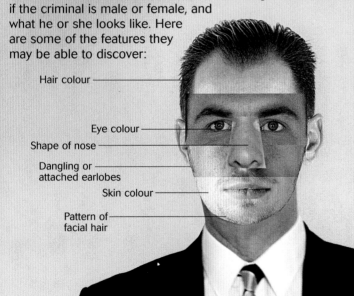

- Hair colour
- Eye colour
- Shape of nose
- Dangling or attached earlobes
- Skin colour
- Pattern of facial hair

Here to stay

So, will gene science make things better or worse on our planet? Only time will tell. But we do know one thing – it's too late to turn back the clock. Cloning, genetic engineering, GM foods and designer babies have been invented, and they can't be uninvented. Over time, they will probably become much more widespread, and much less controversial, than they are today.

In their place will come brand-new genetic developments and inventions – things we probably can't even begin to imagine today. The only thing we can be sure of is that, as has been the case throughout history, there are big changes ahead.

Timeline

This timeline shows the main events in the history of genetics. BC stands for Before Christ, and AD stands for *Anno Domini*, which means "Year of the Lord". The letter c. stands for *circa*, which means "about".

In the 17th century, people used to think sperm cells contained tiny humans, as this drawing shows.

c. 10,000BC Selective breeding of wheat plants began in the area around the eastern Mediterranean.

c. 400BC Hippocrates, a Greek doctor, said that qualities were passed from parents to children in fluids that mixed together, giving a combination of both parents' features.

AD 100-1000 The Hindus noticed that certain body features and diseases could run in families.

1100-1600 Europeans developed the (incorrect) theory of spontaneous generation, which said that living things grew out of non-living matter.

1630 William Harvey realized that babies were made when an egg and a sperm joined (although this had not yet been seen through a microscope).

1856-68 Austrian monk Gregor Mendel studied pea plants, and discovered dominant and recessive genes (which he called "factors"). However, his work was ignored.

1859 Charles Darwin published his book *On the Origin of Species by Means of Natural Selection*. It argued that tiny differences (now known as genetic mutations) allowed species of living things to change over time.

1869 Johann Meischer extracted DNA from white blood cells, though no one yet knew what it was.

1870-90 Using new microscope technology, scientists observed chromosomes and saw cells dividing.

1900 De Vries, von Tschermak and Correns rediscovered Mendel's theories and proved he was right.

1902 The term "gene" began to be used to describe Mendel's "factors".

1905 Edmund Wilson and Nettie Stevens separately discovered that X and Y chromosomes determine whether a person is male or female.

1941 George Beadle and Edward Tatum found that each gene acted as a code for a particular protein.

1944 Oswald Avery and his colleagues found out that DNA carried genetic information.

1950 Erwin Chagaff found that DNA contained equal amounts of the four base chemicals adenine, cytosine, guanine and thymine (often known as A, C, G and T).

1952 Rosalind Franklin studied DNA using X-ray crystallography, and discovered its spiral shape.

1953 James Watson and Francis Crick discovered the molecular structure of DNA.

1956 Francis Crick and George Gamov found out how the bases in DNA coded for different proteins.

1972 Paul Berg modified DNA by splicing two DNA strands together.

1973 Stanley Cohen, Annie Chang and Herbert Boyer combined DNA from two species of bacteria to make the first genetically modified organism.

1975 Fred Sanger and other scientists developed methods for reading DNA sequences.

1977 The company Genentech became the first to make proteins using genetically modified bacteria.

1981 Scientists began to discover genes that contributed to particular illnesses, such as cancer.

1985 Kary Mullis developed the polymerase chain reaction method to copy large amounts of DNA.

1988 Scientists developed the first genetically engineered lab mice.

1989 Alec Jeffreys developed DNA fingerprinting for use in criminal trials.

1990 The Human Genome Project began.

1993 Genetically engineered tomatoes, designed to have an extra-long shelf life, went on sale.

1996 Dolly, the first clone from an adult mammal, was born at the Roslin Institute in Scotland.

2001 The first map of the Human Genome was completed.

2002 Various scientists announced they were working on cloning humans.

2007 Michael Worobey traced the evolutionary origins of HIV, showing that HIV infections had occurred in the US as early as the 1960s.

2014 Nick Tuftnell was saved from blindness by gene therapy in the UK.

Who's who

This who's who lists the most important scientists, thinkers and writers in the world of genes and DNA.

Aristotle (384-322BC)
Ancient Greek scientist and thinker who wrote on many subjects, including biology and genetics. He believed babies inherited all their features from their fathers.

Oswald Avery (1877-1955)
Canadian scientist who specialized in studying bacteria. In 1944 he and his colleagues found that genetic information was carried in DNA.

George Beadle (1903-1989)
American geneticist. In 1941 Beadle and Edward Tatum discovered that each gene acted as a code for a body protein. They were awarded the Nobel Prize in 1958.

Carl Correns (1864-1933) German botanist who in 1900 rediscovered Gregor Mendel's work on genes, and helped prove that Mendel was right.

Francis Crick (1916-2004)
English biochemist who, with James Watson, discovered the molecular structure of DNA in 1953. Along with Watson and Maurice Wilkins, Crick was awarded a Nobel Prize in 1962 for his work on DNA.

Charles Darwin (1809-1882)
English naturalist who spent his life studying living things (after studying medicine and considering becoming a priest). He developed the theory of natural selection, which said that those creatures that were best-suited to their surroundings could survive longer and pass on their qualities to their offspring. This allowed species to evolve, or change, over time.

Richard Dawkins (born 1941)
English zoologist. His books on genes and evolution brought genetics into the public eye in the 1970s.

Rosalind Franklin (1920-1958)
English chemist who developed a way of photographing DNA. This helped uncover the structure of DNA.

Hippocrates (c. 460-370BC)
Ancient Greek doctor. He said male and female substances combined to make a baby with a combination of its mother's and its father's features.

Barbara McClintock (1902-1992)
American geneticist who discovered jumping genes, which move around between chromosomes. She won a Nobel Prize for her work in 1983.

Gregor Mendel (1822-1884)
Austrian monk and scientist who discovered genes (which he called "factors") in the 1860s, by studying pea plants in his monastery garden.

Kary Mullis (born 1944)
American biochemist who developed a method of making lots of copies of DNA. He won a Nobel Prize in 1993.

Fred Sanger (1918-2013)
English chemist who developed DNA sequencing methods in the 1970s. He won Nobel Prizes in 1958 and 1980.

Nettie Stevens (1861-1912)
American biologist who discovered that X and Y chromosomes decide whether an animal is male or female.

John Sulston (born 1942)
English biochemist. He was director of the Sanger Centre, where much of the Human Genome Project was done. He won a Nobel Prize in 2002.

Edward Tatum (1909-1975)
American chemist who, working with George Beadle, found that genes code for body proteins. Tatum and Beadle shared a Nobel Prize in 1958.

Watson and Crick in 1953 with their model of a DNA molecule

Erich von Tschermak (1871-1962)
Austrian agronomist (farming scientist) who in 1900 rediscovered the work of Gregor Mendel.

Craig Venter (born 1946)
American geneticist who developed a new, fast method of reading gene sequences in the 1980s. He became president of Celera Genomics, a human genome mapping company.

Hugo de Vries (1848-1935)
Dutch botanist and naturalist who in 1900 uncovered Mendel's important earlier work on genes. De Vries invented the term "pangene" which was later shortened to "gene".

James Watson (born 1928)
American biologist who worked with Francis Crick to discover the structure of DNA in 1953. He was a Nobel Prize winner in 1962.

Maurice Wilkins (1916-2004) New Zealand-born physicist. He worked with Rosalind Franklin, and shared the 1962 Nobel Prize with Watson and Crick for his contribution to the discovery of the structure of DNA.

Ian Wilmut (born 1944)
English scientist who became the first person to clone a mammal from another adult mammal when he created Dolly the sheep in 1996.

Glossary

This glossary explains some of the difficult or unusual words which you may have seen in this book or in other books about genes and DNA.

A

acid A type of chemical. DNA, vinegar and lemon juice are all types of weak acid.

adenine One of the four bases in DNA that combine in different sequences to make genes.

allele A variation of a gene. For example, genes for eye colour can have blue and brown alleles.

amino acids The 20 chemical building blocks that living things use to build proteins.

ancestors Family members who lived a long time ago, such as your great-great-great-grandparents.

archaeologist A scientist who studies old buildings and human remains to find out about the past.

B

bacteria (singular: bacterium) Tiny living organisms which breed by dividing in two. Some types can cause diseases, but others are harmless or even useful.

base A type of chemical. DNA contains four different bases which combine in different patterns to make up the genetic code.

base pair A set of two bases joined together as part of the structure of DNA. Each base pair forms one "rung" of the spiral-ladder-shaped DNA molecule.

bio-ballistics A way of combining DNA from two species by firing microscopic metal balls, coated with DNA, at living cells. Some of the balls enter the cell nuclei and insert the new DNA into them.

biologist A scientist who studies living things.

bioweapons Weapons designed to hurt or kill their victims by giving them diseases or making them sick. Also called biological weapons.

C

cell The smallest unit of a living thing. Most organisms are made up of many cells.

cell membrane The protective skin surrounding a cell.

chimera A creature from Greek mythology with a lion's head, wings and a serpent's tail. The word chimera can also be used to mean any unnatural combination of different species.

chromosome A strand of DNA found inside a cell nucleus. Most living things have a number of chromosomes in each cell, which together contain a complete set of genes for that organism.

clone A living thing that is an exact genetic copy of another living thing.

cloning Making an identical copy of a living thing, with exactly the same DNA as the original.

cytoplasm A watery or jelly-like substance that makes up most of the inside of a cell.

cytosine One of the four bases in DNA that combine in different sequences to make genes.

D

deoxyribonucleic acid The full scientific name for DNA.

descendants Family members who belong to a later generation, such as great-great-grandchildren.

This picture shows two grains of pollen from a marigold plant, magnified about 1,500 times. Pollen cells are male plant cells. They combine with female cells to make seeds that can grow into new plants.

designer baby A baby grown from an embryo that has been specially selected for its healthy genes.

diabetes A disease in which the body cannot make enough of an important protein called insulin.

DNA (deoxyribonucleic acid) The chemical, found in cell nuclei, that makes up genes and chromosomes.

DNA fingerprinting Comparing samples of DNA to identify someone. For example, DNA from a hair left at a crime scene can be compared to the DNA of suspects.

DNA sequencing Working out the sequence of bases in a sample of DNA. This is usually done by cutting the DNA into small pieces and separating them by passing them through a specially designed gel.

dominant gene The more powerful gene in a pair of alleles. A dominant gene always overrules a weaker recessive gene.

double helix A shape made by two 3D spirals twisting around each other. DNA strands are double-helix-shaped.

dyslexia A medical condition that can make reading, writing and spelling difficult.

E

E. coli A very common species of bacteria which is often used in genetics experiments.

egg A female reproductive cell that can join with a sperm cell from a male to make a new cell that can grow into a baby.

embryo A fertilized egg in the early stages of growing into a baby.

endoplasmic reticula Narrow channels that help transport various substances and molecules around inside a cell.

ethics Guidelines that people use to decide what is right and wrong.

eugenics The science of trying to improve a human population by attempting to control which genes are passed on to future generations.

evolution The gradual change of living things over time.

F

fertilize Make something ready to grow. An egg cell is fertilized when it fuses with a sperm cell, making a new cell that can grow into a baby.

forensic To do with courts of law. Forensic science means using scientific methods, such as DNA testing, to examine criminal evidence.

G

gene A section of DNA in which the bases are arranged in a particular sequence that acts as a code for a particular protein or body substance.

gene mutation A type of error that can happen when a gene is being copied from one cell to another.

gene patenting Taking out a patent (a kind of licence) on a gene that does a particular thing. Genes can be patented by whoever has discovered and studied them.

gene prospecting Taking samples from people or other living things in order to find new genes to patent.

gene therapy Treating genetic diseases by giving patients healthy genes to replace those which are not working properly.

generation A single "level" or step in the history of a species. For example, your parents belong to one generation, and you and any brothers or sisters you have belong to the next.

genethics A word used to describe the ethics of genetic science.

genetic To do with genes.

genetic disease A disease caused, or partly caused, by missing genes or by genes that do not work properly.

genetic engineering Making changes to the genes or DNA of a particular species in order to make it grow and live differently.

genetic modification Another word for genetic engineering. A shortened version, **GM**, is often used to describe genetically engineered crops and farm animals.

genetic trait A feature or quality, such as blue eyes or tallness, that is passed on from one generation to the next in genes.

geneticist A scientist who studies genes and DNA.

genetics The science of genes and DNA.

genome The complete set of genes of a particular species. For example, the human genome is the complete set of all the genes needed to make a human being.

genome mapping Working out the complete sequence of bases in an entire genome.

germ-line therapy Making changes to reproductive cells, such as sperm and egg cells, in order to prevent genetic diseases being passed on from one generation to the next.

GM see **genetic modification**.

GM foods (short for Genetically Modified foods) Foods that come from crops or farm animals whose genes have been modified (changed).

Golgi complex A storage unit inside a cell. The Golgi complex can store spare proteins and send them to wherever they are needed.

guanine One of the four bases in DNA that combine in different sequences to make genes.

H

haemophilia A genetic disease in which a gene for making proteins that help blood to clot is faulty.

helix A 3D spiral shape.

I

immortal Everlasting, or having the ability to live for ever.

insulin An important protein that helps the body to digest sugar.

J

junk DNA Long, seemingly random, repeated sequences of DNA found in between genes.

K

keratin A protein found in skin, hair and nails.

L

ligase A protein used in genetic engineering to help join pieces of DNA together.

lysosome The part of a cell that breaks down and reuses old proteins.

M

mitochondria (singular: mitochondrion) Power units in a cell. They combine food with oxygen to provide energy for the cell's activities.

mitochondrial DNA A small amount of extra DNA stored not in a cell's nucleus, but in its mitochondria.

mitosis The division of a cell into two identical new cells, each with its own nucleus and set of genes.

molecule The smallest particle of a substance that can exist. Molecules (such as the DNA molecule) are made of atoms (tiny units) of different elements joined together.

mummification Covering a dead body in chemicals and wrapping it in bandages in order to preserve it. The ancient Egyptians made mummies that have lasted thousands of years.

mutation see **gene mutation**.

N

nucleus (plural: nuclei) The part of a cell that contains chromosomes and genes. Other parts of the cell use instructions from the genes to do their work. A few types of cells, such as bacteria and red blood cells, do not have their own nucleus.

nurture A name for the upbringing, environment and lifestyle that help to make you who you are. Nurture is often contrasted with "nature", which means the things about you that are decided by your genes.

O

organelles The little "organs", such as ribosomes, lysosomes and mitochondria, that do different jobs inside a cell.

organism A living thing.

ovum Another name for an egg cell.

P

pancreas An organ near your stomach that makes insulin.

patent A legal licence that is used to protect an invention or idea from being stolen or copied. It is now possible to take out a patent on a gene, if you have worked out what the gene does.

PGD see **pre-implantation genetic diagnosis**.

pharming Rearing genetically modified crops or animals to produce useful medicines. The word comes from a combination of "farming" and "pharmaceuticals", which means medicines.

pollen A fine yellow dust released by plants. Pollen grains are the male reproductive cells of plants.

pre-implantation genetic diagnosis (PGD) A method of selecting embryos with healthy genes from a number of embryos grown in a laboratory. The healthy embryos are then implanted in their mother's womb to grow into babies.

protein A chemical made up of amino acids, and found naturally in the bodies of living things.

R

radioactivity A kind of energy released by some substances. It can cause gene mutations, which can lead to diseases such as cancer.

RE see **restriction enzyme**.

recessive gene The weaker gene in a pair of alleles. The instructions carried in a recessive gene are overruled by a dominant gene.

recombinant DNA DNA that contains a combination of genes from two or more organisms.

replicate Make a copy. Often used to describe the way a virus copies itself by invading cells and using them to make copies of itself.

repressor A chemical that attaches itself to DNA to stop a gene from working when it is not needed.

reproduce When a living thing reproduces, it creates more living things of the same kind.

reproductive cells Cells such as eggs, sperm and pollen, which are used to make babies or other offspring when living things reproduce.

restriction enzyme (RE) A type of protein, found in some types of bacteria, that can cut DNA strands in two at particular points.

ribosome Part of a cell that reads the instructions from genes and uses them to make new proteins.

ribonucleic acid see **RNA**.

RNA (ribonucleic acid) A chemical, similar to DNA, that cells use to carry a copy of the genetic code from a gene to a ribosome. Some viruses use RNA instead of DNA to store their genetic code.

RNA polymerase A protein found in cells. It makes an RNA copy of a gene to send to a ribosome.

S

selective breeding Choosing to breed only those plants and animals that have the most useful qualities. Farmers and breeders do this to alter crop and livestock species, such as pigs, horses and wheat, over time to make them more productive and useful to humans.

species The scientific name for a type of plant, animal or other living thing, such as a bacterium.

sperm A male reproductive cell. In humans, a sperm can combine with an egg cell to make a complete cell that can grow into a baby.

stem cells Cells that can grow into any type of body cell.

T

telomerase A protein, found in some types of cells, that can repair the telomeres on chromosomes.

telomere A repeated sequence of bases on the ends of chromosomes. Every time a cell divides, its telomeres wear down, and finally stop it from dividing any more.

terminator seed A type of seed that has been genetically engineered so that the plants that grow from it cannot produce their own seeds.

thymine One of the four bases in DNA that combine in different sequences to make genes.

U

umbilical cord A tube that connects a baby growing in the womb to its mother's body.

V

vector Something that is used in genetic engineering to carry DNA from one organism into another – usually a virus or bacterium.

virus A strand of DNA or RNA in a protective "jacket" of proteins. Viruses can invade cells and use them to make more viruses.

X

X chromosome One of the two chromosomes that determine the sex of a person.

y

y chromosome A type of chromosome passed on by the father that makes a baby into a boy.

Facts and figures

This page contains at-a-glance facts and figures to do with cells, chromosomes, genes and DNA.

Cells

• The average human has between 50 trillion (50,000,000,000,000) and 100 trillion (100,000,000,000,000) body cells at any one time. Every day, over 2 billion (2,000,000,000) cells die and have to be replaced.

• A typical human cell is about 10 microns across – that's 1/100 of a millimetre or 1/2500 of an inch.

• The biggest human cells are egg cells, which grow into babies and are only found in females. They are 100 microns across – just big enough to see without a microscope.

• The longest human cells are the nerve cells which deliver messages from your limbs to your spinal cord. They are very thin but can be up to 1m (over 3ft) long.

• Almost all your body cells have a nucleus containing a complete set of your genes. When cells divide to make new cells, the genes are copied into the new cells.

These are mast cells, one of around 200 different types of cells found in the human body. Mast cells are part of the connective tissue that joins body parts and organs together.

Chromosomes

• Humans have 46 chromosomes inside the nucleus of most of their cells. Chromosomes are long, thin molecules of the chemical DNA.

• You get 23 of your chromosomes from your mum and 23 from your dad.

• Reproductive cells (sperm and egg cells) have only 23 chromosomes each. They can join together to make a whole cell that can become a baby.

• An average chromosome contains about 1,000 genes.

DNA

• DNA stands for deoxyribonucleic acid. It is a long, thin molecule made of a chain of smaller molecules arranged in a double helix shape, which looks like a spiral ladder.

• DNA contains the four bases adenine, cytosine, guanine and thymine. They are arranged in adenine-thymine and guanine-cytosine pairs, known as base pairs.

• The complete human genome has about 3.2 billion base pairs.

• Altogether, the chromosomes in one cell contain about 2m (6ft) of DNA, although it is coiled up to fit inside the nucleus.

• If all the DNA in a human body were stretched out and joined together, it would be up to 200 billion km (120 billion miles) long.

• About 1.5% of our DNA makes proteins. We still don't fully understand what the rest of the DNA is for, though some of it helps to control the genes that make proteins.

• As well as the DNA inside the nucleus, cells have some DNA in their mitochondria (the cell's energy units). This is called mitochondrial DNA and it is not passed on to offspring by both parents, only by the mother.

• All types of living things use DNA to carry genetic information.

• Some viruses use a slightly different chemical called RNA to carry their genetic information. (Scientists disagree about whether viruses count as living things.)

Genes

• The human genome contains about 20,000 genes. Each gene is a sequence of DNA that acts as a code for a body protein or substance.

• Human cells can make more than 250,000 different proteins, because genes sometimes work together in different combinations.

• The average gene is a string of about 1,000 base pairs.

• The largest human gene contains about 2.5 million base pairs. The smallest human gene contains about 500 base pairs.

The gene code

• In a gene, the four bases in DNA form a code which stands for a body substance. The code runs along one side of the DNA strand.

• The four bases are arranged in groups of three.

• Each group stands for an amino acid. Living things use 20 different amino acids in various combinations to make the substances they need.

• The 20 amino acids are: alanine, arginine, asparagine, aspartic acid, cysteine, glutamic acid, glutamine, glycine, histidine, isoleucine, leucine, lysine, methionine, phenylalanine, proline, serine, threonine, tryptophan, tyrosine and valine.

• A whole gene contains a string of three-letter groups that code for a particular string of amino acids. In order to make a protein they need, cells read the right gene and assemble the amino acids in order. The string of amino acids then folds up to form a protein molecule.

Index

A

adenine 9, 12, 56, 58
ageing 42-43
alleles 20, 58
Alzheimer's disease 43
amino acids 12, 13, 16, 56, 58
animals 2, 4, 18, 22, 24, 25, 36,
    45, 46, 50, 51, 52, 54
    cloned 5, 40, 41, 47
    genetically modified 4, 5, 28, 29,
        50, 52, 55, 56
    prehistoric 28, 45
anthrax 33
archaeologists 5, 45, 58
Aristotle 24, 56, 57
Avery, Oswald 27, 56, 57

B

babies 5, 14-15, 18, 19, 22, 24, 25,
    27, 38-39, 40, 41, 43, 45, 47, 54
bacteria 18, 33, 34, 52, 56, 58
    genetically engineered 5, 28, 32,
        33, 36, 55, 56
    E. coli 11, 18, 28, 32, 53, 59
base pairs 9, 13, 31, 58
bases 9, 12, 13, 17, 27, 30, 56, 58
Beadle, George 27, 56, 57
bioweapons 33, 52, 53, 58
blood 8, 9, 12, 16, 22, 36, 44, 56
budding 18, 40
business 29, 50-51

C

cancer 23, 27, 36, 37, 56
cell division 17, 18, 22, 27, 42, 56
cell membrane 8
cell nucleus 6, 8, 9, 16, 17, 26, 60
cells 4, 5, 6, 7, 8, 9, 12, 14, 15, 16,
    17, 18, 22, 26, 28, 31, 32, 34,
    36, 37, 38, 39, 42, 43, 52, 58
    blood 8, 9, 22, 36, 56
    egg 14, 19, 22, 32, 38, 41, 42,
        56, 59
    reproductive 19
    skin 5, 23, 41, 44
    sperm 19, 22, 38, 42, 56, 60
    stem 41, 43
chickens 15, 28
chimera 52, 58

D

chimpanzees 2, 23, 25
chromosomes 8, 9, 10-11, 17, 20-21,
    26, 27, 28, 36, 42, 56, 58, 60
clones 5, 18, 40-41, 58
cloning 5, 40-41, 43, 47, 54, 55, 58
computers 26, 28, 30, 45, 54
copying
    cells 17, 18, 22, 27
    DNA 17, 27, 40
Correns, Carl 26, 56, 57
cotton 33
cows 24, 40, 50, 54
Crick, Francis 27, 56, 57
criminals 5, 28, 44, 46, 55, 56
crops 24, 28
    GM 29, 33, 34-35, 46, 50, 55
cystic fibrosis 21, 37, 38, 49
cytosine 9, 12, 56, 58

dangers 4, 33, 35, 46-47, 48, 52-53
Darwin, Charles 25, 56, 57
databases 45, 51
Dawkins, Richard 57
designer babies 5, 38-39, 48, 49,
    55, 58
diabetes 5, 58
diet 21, 49
disability 47, 48
diseases 5, 16, 20, 21, 24, 28, 29,
    36, 37, 38, 39, 42, 43, 44, 45,
    48, 56
division (of cells) 17, 18, 22, 27, 42,
    56
DNA fingerprinting 5, 44, 45, 56, 58
DNA profiling 55
DNA sequencing 30, 56, 57, 58
DNA testing 5, 28, 36, 38, 44-45
Dolly the sheep 40, 41, 56, 57
dominant genes 21, 25, 56, 58
double helix 9, 27, 58
downloadable pictures 2
dyslexia 49, 58

E

earlobes 20-21
eggs 22, 42, 56
    chicken 15
    human 14, 19, 38, 41, 59
    mouse 32
Einstein, Albert 49
elephants 13
embryos 5, 14, 38, 39, 41, 49, 59
endangered species 54

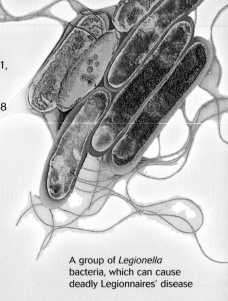

A group of *Legionella*
bacteria, which can cause
deadly Legionnaires' disease

environment, the 35, 43
ethics 29, 33, 39, 40, 41, 46-53, 59
eugenics 48, 59
evolution 22, 25, 59
eye colour 7, 14, 20, 47, 48, 55

F

factor 8 16
famines 35, 47
farming 4, 24, 28, 41
fertilized egg cells 14, 19, 32, 38
fish 29, 34
fleas 26
fetuses 15, 38
"Frankenfoods" 34
Franklin, Rosalind 27, 56, 57
fruit flies 10, 28, 31

G

gel electrophoresis 30, 44
gene patenting 30, 31, 51, 59, 60
gene prospecting 51, 59
gene therapy 29, 37, 50, 59
genetic diseases 5, 16, 20, 21, 24,
    28, 29, 36, 37, 38, 39, 44, 45,
    47, 48, 59
genetic engineering 28, 29, 32-33,
    34, 47, 50, 52, 53, 55, 56, 59
geneticists 24, 28, 59
genetic modification (GM) 4, 5, 33,
    34, 35, 46, 50, 52, 53, 55, 56, 59
genetics 24, 28-29, 31, 59
genetic testing 5, 28, 36, 38, 39,
    44-45, 50

Acknowledgements

The publishers are grateful to the following for permission to reproduce material:
Key: *t = top, m = middle, b = bottom, l = left, r = right*

Cover (DNA molecule) Pasieka/Science Photo Library; cover (DNA sequence) Pasieka/Science Photo Library; title page (viruses) Lee D. Simon/Science Photo Library; contents page (boy & chimp) © Karen Huntt Mason/CORBIS; p4tl (sea lion) Douglas David Seifert/Alamy; p4br (protester) reprinted by kind permission of the Compassion in World Farming Trust, Charles House, 5a Charles Street, Petersfield, Hants GU32 3EH, UK, +44 (0) 1730 268070, www. ciwf.co.uk; p5tl (embryo) Dr Yorgos Nikas/Science Photo Library; p5br (Otzi) © Giansanti Gianni/CORBIS; p6t (octopus) Gary Bell/Alamy; p7tl (skin) Richard Wehr/Custom Medical Stock Photo/Science Photo Library; p7bl (children) © Digital Vision; p8tr (cells) Don Fawcett/Science Photo Library; p8 (cell artwork) Digital Progression; p10ml (leopard) © Theo Allofs/ CORBIS; p10b (cat) © Jane Burton; p10tr (chromosomes) Biophoto Associates/Science Photo Library; p11tr (*E. coli*) Dr Linda Stannard, UCT/Science Photo Library; p11br (chromosome set) Dept. of Clinical Cytogenetics, Addenbrookes Hospital/Science Photo Library; p12tr (alanine) Alfred Pasieka/Science Photo Library; p14tl (fertilized egg) Edelmann/Science Photo Library; p14br (embryo) Edelmann/ Science Photo Library; p15bl (fetus) James Stevenson/Science Photo Library; p15tr (frog) Mark Smith/Science Photo Library; p15br (egg/chick) Hugh Turvey/Science Photo Library; p16tr (hair) Quest/Science Photo Library; p17tr (cells) Stem Jems/Science Photo Library; p18ml (*E. coli* splitting) CNRI/Science Photo Library; p18b (frogs) John M. Burnley/Science Photo Library; p18tr (starfish) © Jeffrey L. Rotman/CORBIS; p19tr (egg & sperm) K. H. Kjeldsen/Science Photo Library; p19br (chess) Oscar Burriel/Science Photo Library; p20bl (young boy) Scott Stulberg/Alamy; p20tr (earlobes) Mark Burnett/Science Photo Library; p21br (violinist) © Ariel Skelley/CORBIS; p22tr (cells) Jackie Lewin, Royal Free Hospital/Science Photo Library; p23tl (lizard) © Joe McDonald/CORBIS; p23tr (cancer) Dr Andrejs Liepins/Science Photo Library; p23b (chimp) Tim Davis/Getty Images; p24tl (Tutankhamun) © Roger Wood/CORBIS; p24br (cave painting) Robert Harding World Imagery/Alamy; p25tl (Darwin) Science Photo Library; p26tm (germ cartoon) Guildhall Library, Corporation of London; p26br (chromosomes) Biophoto Associates/Science Photo Library; p27tl (cell) Dr D. Spector, Peter Arnold Inc./Science Photo Library; p27bm (Franklin photo) Science Photo Library; p27br (DNA) Dr Tim Evans/Science Photo Library; p28tr (fruit fly) Darwin Dale/Science Photo Library; p28bl (GM chickens) © Reuters NewMedia Inc./CORBIS; p29tl (creator) The Art Archive/British Museum/Eileen Tweedy; p29br (salmon) © Natalie Fobes/CORBIS; p30bl (Venter) Hank Morgan/Science Photo Library; p30bl (DNA colonies) Philippe Plailly/Eurelios/Science Photo Library; p30mr (DNA readout) James King-Holmes/Science Photo Library; p31tl (scientist) James King-Holmes/Science Photo Library; p31br (fruit fly) Juergen Berger, Max Planck Institute/Science Photo Library; p32tr (GM *E. coli*) Volker Steger/ Science Photo Library; p32bl (mouse) Eye of Science/Science Photo Library; p33ml (anthrax) © Reuters NewMedia Inc./CORBIS; p33tr (spider web) E. R. Degginger/Science Photo Library; p33br (bollworm) US Dept. of Agriculture ARS/Peggy Greb; p34tm (GM concept) Mehau Kulyk/Science Photo Library; p34bl (monster) © Bettman/CORBIS; p35tl (tomatoes) Volker Steger, Peter Arnold Inc./Science Photo Library; p35tr (butter) Gary Retherford/Science Photo Library; p35bl (protester) © Adrian Arbib/CORBIS; p35br (caterpillar) Marlin E. Rice/Agstock/Science Photo Library; p36bl (tray) Richard T. Nowitz/Science Photo Library; p36tr (bacteria unit) James King-Holmes/Science Photo Library; p36br (chromosomes) Custom Medical Stock Photo/Science Photo Library; p37tr (cancer) Mordun Animal Health Ltd./Science Photo Library; p37b (cystic fibrosis) Juergen Berger, Max Planck Institute/Science Photo Library; p38tl (foetus) Tony Stone Imaging/Getty Images; p38bl (egg cell) Richard Rawlins/Custom Medical Stock Photo/Science Photo Library; p39tl (embryo) Dr Yorgos Nikas/Science Photo Library; p39bl (twins scan) Simon Fraser/Science Photo Library; p39mr (Molly Nash) © Mark Engebretson/CORBIS SYGMA; p40tr (hydra) M. I. Walker/Science Photo Library; p40bl (boys) James Nelson/Getty Images; p41bl (baby) Stevie Grand/Science Photo Library; p41tr (twins) Barbara Penoyar/Getty Images; p41br (Dolly) © Murdo Macleod/CORBIS SYGMA; p42r (bone) Princess Margaret Rose Orthopaedic Hospital/Science Photo Library; p43ml (brain) Alfred Pasieka/Science Photo Library; p43br (scientist) Colin Cuthbert/Science Photo Library; p44tl (DNA) J. C. Revy/Science Photo Library; p45tl (embryo) Dr Yorgos Nikas/Science Photo Library; p45mr (bog body) National Museum, Denmark/Munoz-Yague/Science Photo Library; p45br (spider) © Layne Kennedy/CORBIS; pp46-47 (protest) Karen Robinson/Alamy; p46tr (rat) G. Robert Bishop/Getty Images; p47m (heart) Brian Evans/Science Photo Library; p47tr (baby) © Royalty-Free/CORBIS; p48tl (irises) Ellen Martorelli/Getty Images; p48br (camp survivors) © CORBIS; p49tl (*GATTACA*) The Kobal Collection/ Michaels, Darren; p49bl (pole vaulter) © Jim Cummins/CORBIS; p49br (Einstein) © Bettmann/CORBIS; p50bl (plant) Tek Image/Science Photo Library; p50tr (GM bull) Genpharm International, Peter Arnold Inc./Science Photo Library; p51ml (protein) Dr Tim Evans/Science Photo Library; p51tr (horseshoe crab) Jim Wehtje/Getty Images; p51br (girls) Paul Grebliunas/Getty Images; p52b (virus) Biozentrum, University of Basel/Science Photo Library; p52tr (chimera) © Francesco Venturi/CORBIS; p53tl (*E. coli*) Juergen Berger, Max Planck Institute/Science Photo Library; p53bm (suit) © Leif Skoogfors/CORBIS; p53mr (bomb) Photo Researchers/Science Photo Library; p54b (rhino) © Digital Vision; p54tr (wombs) Victor Habbick Visions/ Science Photo Library; p55tr (Saturn) Space Telescope Science Institute/NASA/Science Photo Library; p55tr (dish) © Lawrence Manning/CORBIS; p55tr (DNA) Will & Deni McIntyre/Science Photo Library; p56 (sperm) Klaus Guldbrandsen/Science Photo Library; p57 (Watson & Crick) A. Barrington Brown/Science Photo Library; p58 (pollen) Dr Jeremy Burgess/Science Photo Library; p61 (cells) Stem Jems/Science Photo Library; p62 (*Legionella*) Dr Linda Stannard, UCT/Science Photo Library; p64 (chromosome) © imagingbody.com

Art director: Mary Cartwright • Picture research by Ruth King
Additional digital image manipulation by Isaac Quaye • Cover design by Tom Lalonde
Special thanks to Paul Dowswell, Susie McCaffrey, Fiona Chandler and Sam Taplin

Every effort has been made to trace and acknowledge ownership of copyright. If any rights have been omitted, the publishers offer to rectify this in any subsequent editions following notification.